# Spin Dependent Transport in Magnetic Nanostructures

Advances in Condensed Matter Science
A series edited by D. D. Sarma, G. Kotliar and Y. Tokura

# Spin Dependent Transport in Magnetic Nanostructures

*Edited by*

## Sadamichi Maekawa
*Institute for Materials Research, Tohoku University, Japan*

## and

## Teruya Shinjo
*Institute for Chemical Research, Kyoto University, Japan*

CRC Press
Taylor & Francis Group
Boca Raton London New York

CRC Press is an imprint of the
Taylor & Francis Group, an **informa** business

CRC Press
Taylor & Francis Group
6000 Broken Sound Parkway NW, Suite 300
Boca Raton, FL 33487-2742

First issued in paperback 2019

© 2002 by Taylor & Francis Group, LLC
CRC Press is an imprint of Taylor & Francis Group, an Informa business

No claim to original U.S. Government works

ISBN-13: 978-0-415-27226-1 (hbk)
ISBN-13: 978-0-367-39602-2 (pbk)

**Library of Congress Cataloging-in-Publication Data**

Catalog record is available from the Library of Congress

**Visit the Taylor & Francis Web site at
http://www.taylorandfrancis.com**

**and the CRC Press Web site at
http://www.crcpress.com**

# CONTENTS

## 3   Experiments of Tunnel Magnetoresistance                           113
TERUNOBU MIYAZAKI

## 4   Theory of Tunnel Magnetoresistance                                 143
SADAMICHI MAEKAWA, SABURO TAKAHASHI AND HIROSHI IMAMURA

# CONTRIBUTORS

**Hiroshi Imamura** Graduate School of Information Sciences, Tohoku University, Sendai 980–8579, Japan

**Peter M. Levy** Department of Physics, New York University, 2 Washington Place, New York, NY 10003, U.S.A.

**Sadamichi Maekawa** Institute for Materials Research, Tohoku University, Sendai 980–8577, Japan

**Ingrid Mertig** Martin-Luther-Universität Halle, Fachbereich Physik, D-06099 Halle, Germany

**Terunobu Miyazaki** Department of Applied Physics, Tohoku University, Sendai 980–8579, Japan

**Stuart S. P. Parkin** IBM Almaden Research Center, 650 Harry Road, K11-D2, San Jose, CA95150–6099, U. S. A.

**Teruya Shinjo** Institute for Chemical Research, Kyoto University, Uji, Kyoto 611–0011, Japan

**Saburo Takahashi** Institute for Materials Research, Tohoku University, Sendai 980–8577, Japan

# PREFACE

The study of electrical transport in magnetic materials has a long history. How-ever, since the discovery in 1988 of the so-called giant magnetoresistance (GMR) effect in magnetic multilayers, the subject has generated a great deal of interest and yielded many surprises. This discovery raised the curtain for an intensive new research effort on magnetic materials and magnetic thin films which profited enormously from the application of microfabrication techniques. Semiconductor devices have been the main actors on the electronics stage during the latter half of twentieth century. The size of such devices has been getting steadily smaller following the impressive progress in microfabrication techniques. In addition, the physics of such small semiconductor devices, i.e., mesoscopic physics, has emerged as a subject of study in its own right. It is also relevant to remember the considerable quantity of new physics which has emerged since the discovery of high temperature superconductors in 1986. In this latter field, one of the main concerns is the behavior of spin and charge and their interactions. On the other hand, to date in semiconductor devices only the movement of charge has found application with spin being usually considered as an irrelevant degree of freedom. The discovery of GMR signaled a starting point for the physics and application of the interaction between spin and charges and shed light to the spin polarized tunneling which was studied in advance of GMR. This book is intended to provide an introduction and guide to the new physics and technology of spin dependent transport in magnetic nanostructures. The emphasis is on magnetic multilayers and magnetic tunnel junctions. The experiment and theory of GMR in magnetic multilayers are described in Chapter 1 (T. Shinjo) and Chapter 2 (P. M. Levy and I. Mertig), respectively. In Chapter 3 (T. Miyazaki) and Chapter 4 (S. Maekawa, S. Takahashi and H. Imamura) are presented the experiment and theory of the tunnel magnetoresistance (TMR). A variety of new TMR problems in magnetic nanostructures such as scanning tunneling microscopy and magnetic nanowires are also discussed in Chapter 4.The applications of GMR and TMR in new technologies and their perspectives are analyzed in Chapter 5 (S. S. P. Parkin). Although the Chapters make up coherently a whole, each Chapter is self-con-tained and may be read independently. The field of magnetic nanostructures is growing rapidly. Therefore, we tried to include the most recent results up to

September 2000. Unfortunately, new developments involving spin transport in magnetic semiconductors which have potential applications for quantum computers occurred too recently to be included. We hope this book is a sound guide to this new physics and technology. Finally, the Editors wish to thank all the contributors in particular the extra effort required to develop such pedagogical reviews.

Sadamichi Maekawa and Teruya Shinjo

# Chapter 1

# Experiments of Giant Magnetoresistance

*Teruya Shinjo*

## 1.1 Introduction

The technical term, magnetoresistance (MR) has been used to express all kinds of electric conductivity change caused by applying a magnetic field and covers a great variety of phenomena. For instance, the features of MR effect in metallic systems are greatly different from those in semiconductors. In ferromagnetic metals and alloys, the difference of resistivity regarding to the direction of magnetization has been known as the anisotropic magnetoresistance (AMR) from a long time ago. Normally, the resistivity is smaller if the electric current direction is perpendicular to the direction of magnetization than in the condition that those are parallel. The origin of AMR is considered to be the orbital angular momentum of magnetic ions and also the Lorentz force acting on conduction electrons. The relative change of resistivity (MR ratio) due to the AMR effect is fairly small; a few present for $Ni_{80}Fe_{20}$ alloy (permalloy) at room temperature and somewhat larger at lower temperatures. Nevertheless, this phenomenon has a significant importance for technical applications such as magnetic sensors. As illustrated in Fig. 1.1, if a magnet is attached on a rotating disk, an MR sensor can detect the number of rotations or the speed of motion very easily from the resistance change of the MR sensor. It has also been attempted to apply an AMR sensor for a magnetic recording technology. Using an MR read-out head, information recorded in a magnetic storage medium is directly converted into electric signals. For ultrahigh density recording, a very high sensitivity of head is indispensable and thus the MR ratio is required as high as possible. However, it was unlikely to find any material having a large MR effect at room temperature. Some magnetic semiconductors have been known to exhibit large MR ratios but only at low

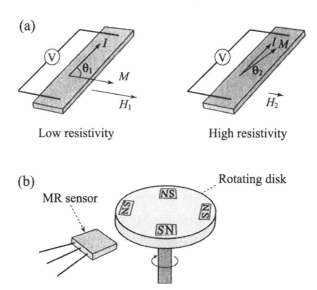

Figure 1.1: (a) Anisotropic magnetoresistance (AMR). $V$, $I$, $M$, $H$ and $\theta$ are voltage, current, magnetization, external field and the angle between the current and the magnetization, respectively. (b) Application of AMR sensor.

temperatures. Although a large MR effect at room temperature under a moderate magnetic field is desirable from a technical viewpoint, no major success has been achieved in the exploration of new MR materials, until the discovery of the giant magnetoresistance (GMR) effect.

The GMR effect has been observed in 1988 by Baibich *et al.* in the resistivity measurements on Fe/Cr multilayers [1], as shown in Fig. 1.2. The discovery of GMR was a great breakthrough in the field of thin film magnetism and magneto-transport studies. At 4.2 K, the resistivity of [Fe 30 Å/Cr 9 Å] multilayer was decreased by almost 50% by applying an external field. Even at 300 K, the decrease of resistivity reaches to 17%, which is significantly larger than MR changes caused by the AMR effect, and therefore the new phenomenon was named "giant". The mechanism of GMR was promptly attributed to the change of the magnetic structure induced by an external field. At zero field, magnetizations in each Fe layer are aligned antiparallel, but are oriented to be parallel by applying an external field larger than 20 kOe. In the antiparallel (AP) magnetic structure, conduction electrons are much more scattered than in a parallel magnetic structure. Until the discovery of GMR, it was not realized that the spin-dependent scattering can make such a large contribution to the resistivity. In the next section, the experimental aspects of GMR effect will be described and studies done before the GMR discovery also are briefly surveyed.

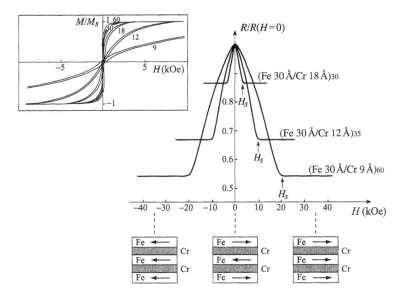

Figure 1.2: Resistivity of Fe/Cr multilayers at 4.2 K as a function of magnetic field. Magnetization curves at 4.2 K are shown in the inserted figure. The numbers represent the Cr layer thicknesses [1].

It is easily speculated that the resistivity of a magnetic material is influenced by spin structures. In a textbook on magnetism, we are able to find for instance the temperature dependence of resistivity of pure Ni metal as shown in Fig. 1.3, which was measured more than 50 years ago. At the Curie temperature of 631 K, the resistivity curve shows a change of the temperature coefficient. The difference of gradients in ferromagnetic and paramagnetic regions is a clear evidence that the resistivity depends on the magnetic structure and the ferromagnetically ordered state has a smaller resistivity. In a paramagnetic region with randomly oriented (or fluctuating) spins, conduction electrons are more scattered by magnetic origins. Indeed, to measure the temperature dependence of resistivity has been known as a method to determine the magnetic transition temperature. However, if you measure the temperature dependence of resistivity for the purpose to determine the transition temperature, you will notice that this method is not sensitive and get an impression that the contribution of magnetism on resistivity is fairly small, because a change of resistivity at the transition temperature is not remarkable. On the other hand, you can extend the resistivity curve in the paramagnetic region (higher than $T_c$) down to lower temperatures without any theoretical guidance (the dotted line in the Fig. 1.3). Then, you will be aware how big is the contribution of the

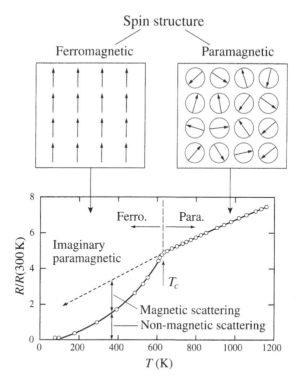

Figure 1.3: Resistivity vs temperature of pure Ni metal. $T_c$ indicates the Curie temperature. The dotted curve is an extension of the resistivity curve in the paramagnetic region to the ferromagnetic region for comparison.

magnetic structure to the resistivity. At room temperature, for instance, the extrapolated resistivity is twice as large as the real value. This comparison is of course non-sense because a magnetically disordered structure at room temperature is fictitious, but is useful to draw an image on the size of spin dependent scattering. From this classic result, one could have got an idea that a large MR effect may happen if the magnetic structure can be varied. However, at lower temperatures than $T_c$, it is generally impossible in normal ferromagnetic materials to realize any kind of disordered spin structure, by applying a moderate magnetic field. The GMR experiment has revealed that the magnetic structure in multilayers composed of ferromagnetic and non-magnetic layers can be varied by applying external fields. We have obtained a hint here how to construct multilayers whose magnetic structures may be modified by external magnetic fields.

As is described in Chapter 2, the GMR effect has evidenced that the spin dependent scattering can induce a significantly large MR change at room temperature. Subsequently, studies on various phenomena relating to the interplay of magnetism and transport have been very much stimulated in fundamental and also technological aspects, and the term "magnetoresistance" has become a fashionable key word. Besides metallic multilayer systems, similar phenomena are found in granular systems where small ferromagnetic clusters are dispersed in non-magnetic matrices. Extremely large MR effect has been found in manganese perovskite oxides and is named to be the colossal MR (CMR) effect which means to be greater than giant. Latest achievements on perovskite oxides have been described in another volume "Colossal Magnetoresistive Oxides" edited by Tokura in the same monograph series as this. Currently, remarkable progress has been made also in the studies on MR effect using tunneling currents, which is often abbreviated to be tunnel magnetoresistance (TMR). Detailed description on TMR will be given in other chapters. In order to express more comprehensively the GMR including related phenomena such as CMR and TMR, we may denote as "XMR" (X = G, T, C). If necessary, X may include A (i.e., AMR) also. The term, XMR can cover all novel MR effects and in addition X sounds to express our further wish to find something new and exotic.

## 1.2 GMR in multilayered systems: experimental aspects

### 1.2.1 Before the discovery of GMR

Pioneering works to study the dependence of conductivity on the magnetic structure have been carried out for tunneling junctions much earlier than the discovery of GMR. A tunneling current from one ferromagnetic metal to another ferromagnetic one through a thin potential barrier may depend on the relative orientations of the two magnets. This idea was verified for the junctions of Co/Ge/Ni and Ni/NiO/Co respectively by Julliere [2], and by Maekawa and Gäfvert [3]. However, they have observed only small MR effect at low temperatures and the results did not gather intense attention at that time. In very recent years, there have been remarkable progresses in the studies of tunneling current MR (TMR), experimentally and theoretically. Since it has been confirmed that a large MR ratio can be achieved at room temperature, the importance of TMR also for industrial sides is recognized, as will be mentioned in the forthcoming chapters.

In ferromagnetic thin films, the magnetic structures become more or less inhomogeneous in the process of magnetization reversal and disordered magnetic fractions, domain walls for instance, will act as sources of resistance. In a resistivity measurement as a function of external field, it is common to observe an increase (or a decrease) of resistivity around the field for the magnetization reversal. An increase is attributed to the spin-dependent scattering and a decrease, AMR effect. However, in any case, variations of resistance are not very large. Magnetic and electric behaviors of a sandwich system consisting of ferromagnetic/non-magnetic metal layers were studied by Dupas *et al.* [4]. They prepared epitaxially grown 3 Å Co layer sandwiched by two 300 Å Au layers, [Au 300 Å/Co 3 Å/Au 300 Å], and measured the resistivity with sweeping the magnetic field. The resistivity showed a clear increase at the field of magnetization reversal such as 1% at 300 K, and 6% at 4.2 K. The observed enhancement of resistance is caused by the magnetic structure change of monoatomic Co layer. As a magnitude of MR ratio, this value is not remarkably large, but if we consider the thickness of the Au layers being 100 times larger than that of Co, the contribution of magnetism to the total conductivity appears to be significant. Whether this phenomenon is related to the preferred orientation of magnetization being perpendicular to the film plane or not still open question. The resistance should depend on the magnetic structure from microscopic viewpoints and therefore the perpendicular anisotropy may play a crucial role.

The study on the Fe/Cr/Fe sandwich system has been initiated by Grünberg *et al.* [5]. This is a preceding work very closely relating to the GMR studies by Baibich *et al.* [1]. In 1986, Grünberg's group found that an antiferromagnetic exchange interaction exists between Fe layers separated by a Cr 10 Å layer by measuring the hysteresis curves utilizing magneto-optic Kerr effect (MOKE) and light scattering (LS) due to spin waves. Spin polarized LEED measurements by Alvarado and Carbone also confirmed the existence of antiferromagnetic interlayer coupling [6]. In a trilayer sample, [Fe 120 Å/Cr 10 Å/Fe 120 Å], two Fe layers' magnetizations are oriented antiparallel in the absence of external field, because of the interlayer exchange coupling, but reoriented to be parallel by applying an external field larger than 3 kOe. Grünberg *et al.* have also measured the resistivity of Fe/Cr/Fe sandwich film with applying external field and observed the difference of resistivity between the states of parallel and antiparallel magnetizations. It has been clearly evidenced that the conductance is influenced by the magnetic structure due to the spin dependent scattering of conduction electrons [7]. Thus, the physical principle of GMR was already demonstrated in this experiment. However, the observed MR ratio, about 1.5% at low temperatures, was not so large to gather attention widely as a discovery of new MR effect.

In the research on metallic magnetic systems, one of major problems has been the long range exchange interactions acting between two local magnetic impurities. The Ruderman-Kittel-Kasuya-Yosida (RKKY) interaction has been known to be the mechanism to account for long range interactions through conduction electrons. However, interactions between impurities are rather weak, and therefore were the subjects in the field of low temperature physics. Until the Fe/Cr/Fe studies, exchange interactions between magnetic layers with a distance as 10 Å was thought to be fairly weak. Although already many kinds of multilayers consisting of magnetic and non-magnetic components had been studied, we have hardly imaged such a magnetic structure that magnetizations in adjacent layers are oriented entirely antiparallel by the interlayer coupling [8]. Nearly at the same time with the Fe/Cr/Fe studies, but independently, Cebollada *et al.* confirmed by neutron diffraction technique that an antiparallel magnetic structure is realized in Co/Cu multilayers [9]. This is the first confirmation of the antiparallel alignment of magnetizations in a multilayer systems.

## 1.2.2 The GMR effect (Fe/Cr multilayers)

Baibich *et al.* prepared Fe/Cr multilayered films by molecular beam epitaxy (MBE) technique initially to study in more details the magnetic behaviors of the multilayer system with antiferromagnetic interlayer coupling. The samples with a structure for instance, [Fe 30 Å/Cr 9 Å] × 60, were deposited on GaAs (001) substrates and the epitaxial stacking of Fe(001)/Cr(001) was confirmed [1]. The magnetization curves for samples with various Cr layer thicknesses are also shown in Fig. 1.2, in the preceding chapter. By applying external field, the resistivity has decreased greatly. At the saturation field, the resistivity is almost a half of that at zero field, which was an unexpectedly large MR effect for metallic substances. The mechanism of the GMR effect was phenomenologically interpreted by considering the spin-dependent scattering of conduction electrons. It is assumed that the scattering probability depends on the relation between conduction electrons' spin direction, up or down, and the directions of magnetization in magnetic layers. For instance, an up-spin electron is considered to penetrate without scattering from Cr layer into Fe layer having magnetization with the up-spin direction, while a down-spin electron is scattered. If Fe layers' magnetizations are antiparallel, both up- and down-spin electrons will meet an Fe layer with magnetization of opposite direction soon (within two Fe layers' distance) and accordingly have a rather high possibility to be scattered. On the other hand, if all the Fe layers' magnetizations are parallel, down-spin electrons are scattered at every Fe layer while up-spin electrons are not, having a long mean-free path. As a total of conductivities due to up-spin and down-spin currents, the state with parallel magnetizations has a much better conductivity than that with antiparallel magnetizations. Explanation of spin-dependent scattering from theoretical aspects is given in the theoretical chapter of this volume.

From the discovery of the GMR effect, we have learnt the following: (1) Spin-dependence scattering can make a significantly large contribution to the total resistance. (2) An interlayer coupling exists between ferromagnetic layers separated by a spacing layer, which can be fairly strong. The magnetization measurements shown in Fig. 1.2 suggest the existence of magnetic structure

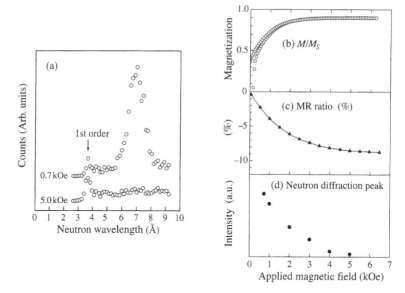

Figure 1.4: (a) Time of flight polarized neutron diffraction profiles for the 1/2-order peak from a multilayer, [Fe 2 Å/Cr 12 Å]×30 at 300 K [10]. The 1st order peak is also observed. The relative intensities of 1st and 1/2-order peaks are not proportional to the structure factors because of the wavelength dependence of the incident neutron beam intensity. External field dependence of magnetization, resistivity, and the 1/2-order neutron diffraction peak intensity at 300 K of a multilayer [Fe 27 Å/Cr 12 Å]×30 is shown as Figs. 1.4(b), (c) and (d), respectively.

with antiparallel alignment. The external field for saturation increases with the decrease of the Cr layer thickness. The sample with 9 Å Cr layers shows very slow increase of magnetization when the external field increases. The antiferromagnetic coupling between adjacent ferromagnetic Fe layers seems to be very strong. A more direct evidence of the antiparallel alignment of Fe layers' magnetizations was given by observing a neutron diffraction peak due to the antiparallel superstructure [10]. The neutron diffraction peak of an Fe/Cr multilayer, shown in Fig. 1.4, corresponds to the twice of the adjacent Fe layer distance, and therefore clearly indicates the antiparallel alignment of Fe layer magnetizations. The intensity of the peak decreases with the increase of external field and it disappears at around 4 kOe. The field dependence of the peak intensity is shown in the figure, together with the change of the magnetization and resistivity. It is apparent from the comparison that the GMR effect in Fe/Cr multilayers is caused by the antiparallel alignment of magnetic layers.

In general, it is possible to observe a GMR-type resistivity difference not only between parallel and antiparallel magnetic states, but also between ferromagnetically aligned states and any disordered states. The GMR effect in granular systems to be mentioned later is the difference of resistivity between the parallel and the randomized spin structures. As for spin structure, an antiparallel alignment is one of the perfectly ordered state but concerning the spin dependent scattering for conduction electrons, an antiparallel state is equivalent to a disordered state. The result on Fe/Cr multilayers has revealed that a significantly large MR effect, which may be much larger than the classic AMR effect, is realized if the magnetic spin structure is able to be varied as desired. However, in the case of Fe/Cr multilayers, the externally applied fields are rather large, in the order of 1 kOe or more, in order to overcome the antiferromagnetic interlayer coupling. Therefore, at the initial stage the potential of the GMR effect for technical applications did not appear to be very promising. In the initial report on the GMR effect in Fe/Cr multilayers by Baibich *et al.* [1], the resistivity at zero field is defined to be 100% and the decrease of resistivity is expressed relative to the value at zero field. The resistivity becomes a constant if the magnetization has saturated. Here, all magnetizations are regarded to be entirely parallel and the spin-dependent scattering due to disordered magnetic structures does not exist. The resistivity at the saturated state should be regarded as a basic value where the contribution from spin disorder is minimized. Therefore, it makes sense to express the resistivity relative to the value at the saturated region and after a while, it has become common to express the GMR ratio relative to the value at saturation, instead of that at zero field, i.e.,

$$\text{MR ratio (\%)} = \frac{[R(H) - R(\text{saturation})] \times 100}{R(\text{saturation})} \qquad (1.1)$$

where $R(H)$ means the resistivity under external field, $H$. In this manner, the MR ratios of the Fe/Cr multilayer at 4.2 K and 300 K in the original report have turned out to be 85% and 19%, respectively. At lower temperatures, the resistivity of metals due to electron-phonon interaction should decrease. In contrast, the contribution due to the spin-dependent scattering does not decrease at low temperatures. In the case of GMR effect, an MR ratio means the relative size of magnetic resistance to the total. Therefore, the MR ratio increases with a decrease of temperature. In a later report, the MR ratio of Fe/Cr multilayers at very low temperature has attained to 200% [11]. Nowadays almost all MR results are shown relative to the saturated value but if the saturation field is very large, the resistivity at the saturated state is not easily estimated experimentally and for the convenience the resistance change is expressed relative to the zero field value. Such examples, having very large saturation fields, are often seen in granular systems.

Since the discovery of GMR effect, interlayer couplings between ferromagnetic layers across a non-magnetic intervening layer has gathered attention and various studies have been intensely carried out even independently of the interest in GMR. As will be described in Sec. 1.2.3, the role of non-magnetic spacer layer is crucial to account for the interlayer coupling between ferromagnetic layers. In the last decade, many experimental and theoretical studies have been performed for magnetic/non-magnetic/magnetic systems, primarily for the purpose to investigate the condition to enhance the MR ratio [12]. Consequently, the GMR discovery has played as a great breakthrough and the influence has extended widely to various fields. Our understanding of the electronic structure in ultrathin metal films has greatly progressed and for instance the existence of quantum well states in metallic layers have been elucidated [13].

## 1.2.3  Other coupled-type GMR systems and interlayer coupling

Studies on multilayers consisting of magnetic and non-magnetic layers have been stimulated by the GMR effect found in Fe/Cr multilayers. The existence of interlayer coupling was confirmed in many systems and it has been observed that an antiparallel magnetic alignment generally has a larger resistivity than a parallel (ferromagnetically aligned) state. However, the MR ratio is not always very large but depends on the combinations of magnetic and non-magnetic substances. Among magnetic/non-magnetic multilayers, a particularly interesting system is Co/Cu multilayers, of which MR behaviors were initially reported by two groups and succeedingly examined by many investigators [14, 15]. The MR ratio of Co/Cu multilayers is very large as well as that of Fe/Cr systems and the coercive force is smaller than the case of Fe/Cr. Since Cu is a typical non-magnetic metal, the understanding of the mechanism of interlayer coupling across a Cu spacer layer appears to be simpler than that for Cr spacer layer. Practically, Cu-based multilayers show larger MR ratios than other noble metals and therefore Cu has been adopted as a spacer material the most frequently. Concerning interlayer couplings, a striking discovery was the oscillatory dependence of the interlayer coupling strength on the spacer layer thickness, or in other words the distance between ferromagnetic layers [16]. In Fig. 1.5, the MR ratio of Co/Cu multilayers is shown as a function of Cu layer thickness. The peaks of MR ratio appear discretely but periodically. The saturation field also becomes the largest at the peak position of MR ratio, suggesting that the MR effect is relating to an antiferromagnetic interlayer coupling and a strong coupling appears at some specific layer thicknesses of the spacer.

Parkin *et al.* have prepared various multilayer samples of Co in the combination with other metal elements by sputtering technique and found the oscillation of interlayer coupling with the wavelength of $10 \sim 15\,\text{Å}$ occurs rather

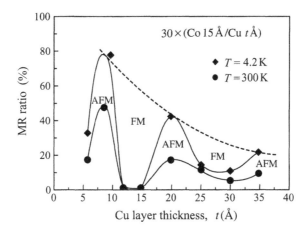

Figure 1.5: MR ratio of Cu/Co multilayers as a function of Cu layer thickness [14].

generally [17]. Problems of interlayer couplings have then gathered great attention and very intense investigations have been carried out in a short period, to elucidate the role of non-magnetic spacer layers.

For the study of interlayer coupling problems, trilayer (sandwich) samples consisting of magnetic/non-magnetic/magnetic structures are useful, which have a physically almost equivalent with magnetic/non-magnetic multilayer samples. The preparation of trilayer samples is simpler than that of multilayers and the quality of surface (or interface) in trilayer can be higher than in multilayers. If a sandwich samples with a "wedge" structure is prepared, the thickness dependence of magnetic behaviors is able to be studied systematically by using one sample. Examples of wedge structure are shown in Fig. 1.6. A simple wedge structure as shown in Fig. 1.6(a), has a non-magnetic spacer layer with a gradually varying thickness for instance by 1 nm in the distance of 1 mm. By applying spin-polarized scanning electron microscopy (SEMPA) or Kerr magneto-optic techniques, the direction of local magnetization is measured [18]. SEMPA measurements on Fe/Cr-wedge/Fe trilayer grown on a whisker substrate showed clearly the periodic change of the direction of magnetization, which corresponds to the oscillatory change of the interlayer coupling. Magnetic hysteresis curves for each location and magnetic behaviors are studied systematically as a function of thickness.

Using a wedge sample, normally it is very difficult to measure the resistivity as a function of layer thickness but rather easy to measure the local magnetization curves. From the variation of saturation field, the oscillation of interlayer coupling as a function of layer thickness is estimated. A double wedge struc-

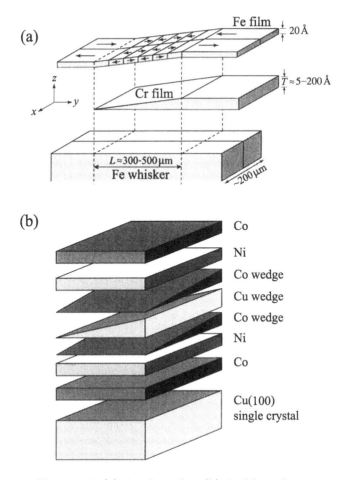

Figure 1.6: (a) simple wedge, (b) double wedges.

tures shown in Fig. 1.6(b) is useful to study the dependence on magnetic layer thickness simultaneously with that on non-magnetic layer thickness [19].

   If the spacer material is a noble metal, the oscillatory behavior of inter-layer coupling through a spacer layer is understood as due to the formation of standing waves (quantum well states) in the spacer layer. The most popularly studied case is Fe/Au/Fe in which interlayer coupling oscillates with two wave-lengths, long and short. Since the Fermi surface structure is known for bulk Au metal, the possible wavelength of standing wave along the direction perpen-dicular to the film is speculated and experimentally observed wavelengths are in a fairly good agreement with the theoretical prediction [20]. After the dis-covery of interlayer coupling oscillation, strong attention has been paid on the electronic structure of metallic ultrathin films and the existence of quantum well states in ultrathin Cu and Ag layers has been found by the inverse photo-emission spectroscopy by Himpsel, and the relation between interlayer coupling has been argued [13]. On interlayer coupling studies, several comprehensive review articles have been published [21].

If the spacer layers have their own magnetism, the situation is much more complicated than the case of noble metals. Although Fe/Cr/Fe is the initial system for GMR effect, there are unsettled problems because of the complex magnetic behaviors of Cr metal layers. By recent studies using neutron diffraction [22, 23] and Mössbauer spectroscopy [24], the Néel temperature of ultrathin Cr layers is confirmed to be much higher than room temperature. Therefore, the contribution of the antiferromagnetism in a Cr spacer layer to the interlayer coupling between magnetic layers should be significant. The neutron diffraction studies suggested that the magnetic structure of Cr layers separated by a ferromagnetic layer holds a coherence. If such a long range magnetic coherence exists, the situation of multilayers may be different from that of trilayers. Pierce *et al.* reviewed the results on Fe/Cr/Fe trilayer studies and discussed the difference from multilayers [25].

## 1.2.4   Non-coupled systems (multilayers)

In the multilayer systems with antiferromagnetic interlayer couplings so far described, antiparallel magnetic alignments are converted into ferromagnetic ones by applying an external magnetic field. The GMR means the great difference of conductivity between antiferromagnetic and ferromagnetic states due to spin-dependent scattering of conduction electrons. Similar phenomena take place in non-coupled multilayer systems, where a magnetic structure transformation is not caused by interlayer coupling but by the difference of coercivities. Multilayers including two magnetic components with different coercivities were prepared by stacking NiFe, Cu, Co and Cu, succeedingly [26]. If the thickness of Cu spacer is not too thin, magnetic layers are isolated with each other and two magnetic components show the reversal of each magnetization independently. In Fig. 1.7, the magnetization curve of [NiFe 30 Å/Cu 50 Å/Co 30 Å/Cu 50 Å]×15 at 300 K is reproduced. The curve is the sum of the two magnetizations in NiFe and Co layers. NiFe is a typical soft magnetic material having a small coercivity while Co has a much larger one. The switching fields of NiFe and Co layers are observed to be about 10 and 100 Oe, respectively.

An external field enough to orient both magnetizations in the field direction was first applied and then the magnetization and resistivity was measured with varying external field. If the field exceeds $H_c$(NiFe), the magnetization of NiFe layers turns to the direction of the external field but that of Co layers remains unchanged until $H_c$(Co). Eventually, two magnetizations are expected to be antiparallel between $H_c$(NiFe) and $H_c$(Co). The change of magnetic structure influences the resistivity greatly and a remarkable enhancement of resistivity

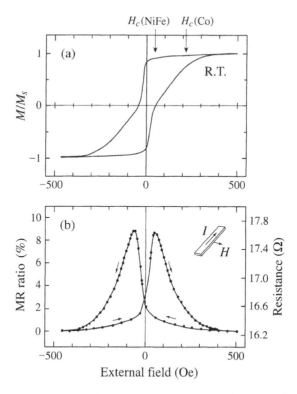

Figure 1.7: Magnetization (a) and resistivity (b) of [NiFe 30 Å/Cu 50 Å/Co 30 Å/Cu 50 Å]×15 at 300 K [26].

is observed in the field region between $H_c$(NiFe) and $H_c$(Co). Apparently, the GMR effect is working also in a non-coupled multilayer system. The resistivity vs field curve shows a rather sharp peak. If both the reversals of two magnetizations occur suddenly, a plateau should have been observed but because the magnetization reversal in the Co layers is gradual, the profile becomes not a plateau but a peak. It is general that a magnetization is gradually reversed with forming domain structures. Afterwards, in many cases, plateau-type MR profiles have been found, for example in wire-shaped systems where the domain wall formation is controlled [27].

In the case of the non-coupled GMR system also, it was confirmed by observing a superstructure peak of neutron diffraction that the majority of magnetic structure is antiparallel [28]. The GMR effect observed here is regarded as the difference of resistivity between a parallel and an antiparallel magnetic state. From this experiment, several suggestions are obtained. For the GMR effect, interlayer couplings are not a necessary condition but magnetic struc-

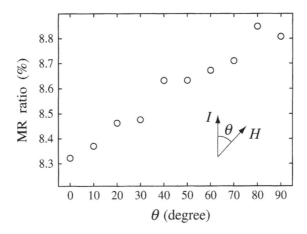

Figure 1.8: Non-coupled type GMR ratio in [NiFe 25 Å/Cu 55 Å/Co 25 Å/Cu 55 Å]×15 as a function of the angle between the magnetic field and the electric current [30].

tural changes by any reason may cause an MR effect. It is of practical importance that the working field is fairly small. In contrast to the coupled systems, where a large field is required to overcome the interlayer coupling, non-coupled systems require only a moderate field. If one component is very soft having a small switching field, the MR effect is exhibited in very weak fields. From the observation of non-coupled type GMR effect, the potential of the GMR effect for technical applications has been recognized to be very promising. Actually, a great success for technical use has been achieved in sandwich system including two magnetic layers, which was named "spin valve" [29]. The description on spin valve systems will be given in the next section.

Non-coupled multilayer systems are useful for various experiments to study the fundamental aspects of GMR phenomena. Antiparallel magnetic structures are easily realized for every case regardless to the thickness of spacer layer, unless it is too thin. In contrast, in coupled systems, the interlayer coupling works strongly only at certain spacings. Because an antiparallel alignment is realized at specified thicknesses, the relation between MR ratio and the spacer layer thickness, for instance, cannot be studied continuously. On the other hand, in non-coupled systems, an antiparallel alignment is realized at any spacer layer thickness. In addition, in the latter case, not only collinear alignments but also non-collinear magnetic structures are realized. An example on non-collinear magnetic structures will be presented in Fig. 1.9. Taking the advantage of non-coupled type systems, several investigations were carried out.

In contrast to the AMR effect, which means the difference of resistivity regarding to the direction of magnetization, the GMR does not depend on the direction of the magnetic field, which is verified experimentally by measuring the GMR ratio of [NiFe/Cu/Co/Cu] as a function of the relative angle between the magnetic field and the current directions. Then the observed MR ratio was found to depend only slightly on the angle (about 0.5% in the present case as shown in Fig. 1.8). Although NiFe layers exhibit a rather large AMR effect, the influence to the total MR behaviors is fairly small since the majority of electric current flows in Cu layers of the [NiFe/Cu/Co/Cu] multilayer and therefore the contribution of the AMR effect in NiFe layers to the total MR effect is not significant.

By utilizing non-coupled type GMR multilayers, it is possible to check experimentally the dependence of GMR on the relative angle between magnetizations in adjacent layers [30]. For the spin-dependent scattering, the antiparallel alignment of magnetizations is the optimum situation and the spin-dependent scattering probability will decrease with $\cos\theta$ where $\theta$ expresses the relative angle between two magnetizations. The MR effect of [NiFe/Cu/Co/Cu] was measured by changing the relative angle between magnetizations in NiFe and Co layers. Namely a moderate external field, which is enough strong to reorient the magnetization of NiFe layers but too weak for the reorientation of Co layers' magnetization, was applied in various directions. Figure 1.9(a) is the resistance of [NiFe 25 Å/Cu 100 Å/Co 25 Å/Cu 100 Å]×15 measured by changing the direction of external field, 20 Oe. In this case, the Cu layer thickness, 100 Å, is sufficiently large to isolate each magnetic layer. After applying an external field in a certain direction for the saturation, the angle dependence was measured. The result is shown as a function of relative angle between the directions of the initial magnetic field and the rotating field of 20 Oe. The direction of Co layers' magnetization should be the same as that of the initial magnetic field while the direction of NiFe layers' magnetization is thought to follow that of rotating field. As shown in the figure, the obtained result is simulated well by a cosine curve as expected. Thus, it is proved that the probability of spin dependent scattering is proportional to $\cos\theta$. More strictly speaking, the obtained curve in Fig. 1.9 shows a minor deviation from an ideal cosine function but is accounted for with assuming a contribution of AMR effect in NiFe layers whose magnitude is 1/60 of the whole MR effect. The result in Fig. 1.9(b) was obtained for a similar sample with thinner Cu layers, 55 Å. The obtained curve has a distorted profile from an ideal cosine function and the reason is thought to be the Cu spacer layer thickness being insufficient for magnetic insulation. If a magnetic interaction from Co layers extends to NiFe layers, the magnetization in NiFe layers does not follow completely the rotating external field. If a magnetic field induced by the magnetization in Co layers is assumed to exist at NiFe layers, the effective magnetic field acting at a NiFe layer should be a vector sum of the external field, 20 Oe, and the stray field. By assuming the stray field in the direction of Co layer's magnetization with various magnitudes, behaviors of resistance as a function of the relative

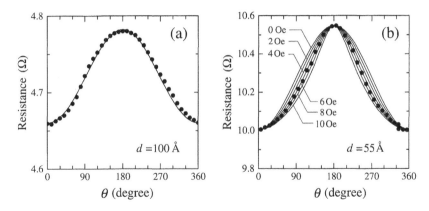

Figure 1.9: MR effect of [Cu($d$)/Co 25 Å/Cu($d$)/NiFe 25 Å]×15 at 300 K under a rotating field of 20 Oe. The angle $\theta$ means the direction of the external field. (a) $d$=100 Å and (b) $d = 55$ Å [31].

angle of external field are expected as the curves in Fig. 1.9(b). Compared with the experimental results (dots in the figure), it is found that 6 ~ 8 Oe gives a reasonably good fitting. This result means that the magnetic influence at NiFe layer from Co layers separated by 55 Å Cu spacer layer corresponds to about 7 Oe. As a magnitude of magnetic field, this value is small but not negligible. The origin should be a sum of several contributions; dipole field, exchange field due to conduction electron polarization and also direct coupling if any pin holes exist. Here presented is an experimental result, which should depend on the sample. It is not certified that the conclusion has a general meaning or not.

In non-coupled type multilayers, the spacer layer has to be enough thick to reduce the interlayer coupling between magnetic layers. Unless magnetic layers are able to switch the direction of magnetization independently, antiparallel magnetizations are not realized and subsequently non-coupled type GMR phenomena do not take place. If the spacer layer is very thick on the other hand, magnetic insulation should be certified but the magnitude of the MR ratio cannot be very large. The MR ratio may be the maximum when the spacer layer thickness is the smallest. The MR ratio as a function of the spacer layer thickness has been checked for three metals, Cu, Ag and Au [31]. These materials are typical novel metals often used as non-magnetic spacer layers. It is shown in Fig. 1.10 that Cu gives the best result. Although the MR ratio in Ag-based films is larger in the region of thick spacer layers, the maximum is not larger than that of Cu-based films because the critical thickness for Ag-based films, where the interlayer coupling appears, is larger than that of Cu-based films. Perhaps this result is not due to the intrinsic electronic structures of the

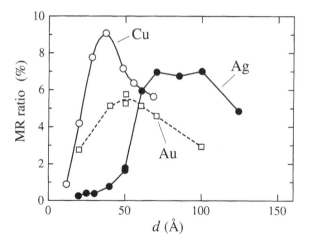

Figure 1.10: Comparison of the non-coupled type GMR effect in Cu-, Au- and Ag-based films. The thicknesses of magnetic layers (NiFe and Co) are 25 Å. The number of multilayer repetition is 5, for all samples [32].

metals but is owing to the quality of actually prepared interfaces. In Ag-based films, a ferromagnetic interaction is observed if the Ag layer thickness is small, which suggests the existence of pin holes. In general, the quality of interfaces can be better in three layer (spin-valve) systems than that of multilayers.

It has been a subject of argument whether the spin-dependent scattering occurs at the interface sites or in the bulk region. This issue is able to be studied experimentally by using samples of non-coupled GMR systems with modified interfaces. The results indicate that the spin-dependent scattering occurs predominantly at interface sites, depending critically on the interface elements. Figure 1.11 shows the magnetization and the MR ratio for non-coupled type GMR multilayers, with and without a Cr-doping layer. The nominal structures are [NiFe 20 Å/Cu 35 Å/Co 10 Å/Cu 35 Å] × 5, and [NiFe 20 Å/Cu 35 Å/Cr 0.5 Å/Co 10 Å/Cr 0.5 Å/Cu 35 Å]×5, respectively. Both were prepared by UHV deposition on glass substrates. Figure 1.11(b) indicates that the magnetization curve has two steps and therefore the interlayer coupling is negligible in both cases. The coercive field of Co layers is decreased by the Cr doping but the overall profile has no change and the establishment of antiparallel magnetic alignment is certified in both cases. The undoped sample exhibits a rather large change of resistance due to non-coupled type GMR effect. On the other hand, the sample with Cr-doped interfaces shows a quite different MR profile. The spin dependent scattering has almost vanished because of the interface doping with extremely thin Cr layer. This result evidences that the role of interface is crucial for spin-dependent scattering.

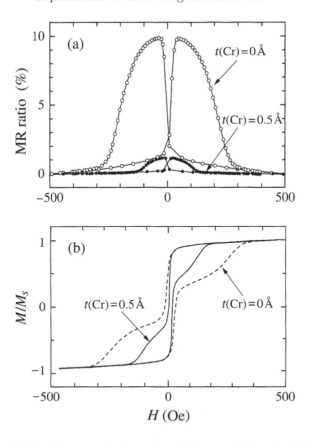

Figure 1.11: MR and magnetization of multilayers, with and without the interface doping by 0.5 Å Cr layer [32].

In the case of doping with Cr, the influence is drastic even if the nominal Cr layer thickness is less than one monolayer. It is of interest to check the influences by the doping with other elements. Similar experiments were made systematically for Cr, Mn, Fe, Co and NiFe. Impurity layers with nominally 1 Å thickness were inserted in all magnetic/non-magnetic interfaces [32]. Namely the nominal multilayer structure is [TM 1 Å/NiFe 20 Å/TM 1 Å/Cu 35 Å/TM 1 Å/Co 10 Å/TM 1 Å/Cu 35 Å]×5, where TM means Cr, Mn, Fe, Co or $Ni_{80}Fe_{20}$. The observed MR ratios at room temperature are shown in Fig. 1.12 together with the variation of the sheet resistivity. When the interface is doped with Cr or Mn layers, the MR ratios decrease drastically. The difference of resistivity due to the magnetic structural change, $\Delta\rho$, is also in the figure, whose variation is very similar to that of the MR ratio. Although the resistivities of Cr- and Mn-doped samples are somewhat larger, the main

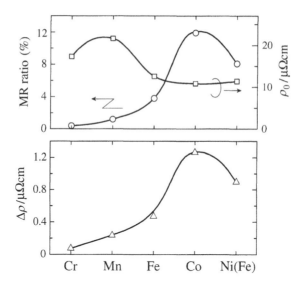

Figure 1.12: Influence of interface doping with 3$d$ element on MR [32].

reason for the reduction of the MR ratio is apparently the disappearance of spin-dependent scattering. The variation of the values on the elements shown in the figure indicates that the maximum of the MR ratio is given by the doping with Co and the MR ratio becomes almost zero for Cr-doping. This tendency is consistent with theoretical predictions on the MR effect of Cu/TM multilayers [33]. It is however impressive that a submonolayer located at an interface plays such a critical role. It is worth to emphasize that GMR properties depend greatly on the situation of interface.

If the mean free path of conduction electrons has varied, not only electric conductivity but also other transport properties may be changed. Magnetic field effect on thermal conductivity associated with the GMR effect has been detected for the first time on a non-coupled type multilayer system by Sato *et al.* [34]. The sample was a multilayer with the structure of Cr 30 Å/[NiFe 20 Å/Cu 40 Å/Co 20 Å/Cu 40 Å] × 50 deposited on a glass substrate with an area of 5 × 10 mm. The field dependence of electric conductivity (namely the GMR) is shown for comparison. The set-up to measure the thermal conductivity is illustrated in the figure. In order to measure the field dependence of thermal conductivity, the temperature gradient along the sample was monitored continuously using Au(Fe) thermocouple. One end of the sample was connected to a Cu heat sink while a small heater was attached on the other end to make a certain temperature gradient. If the heat flow is constant, the temperature difference measured by the thermocouple should be a constant. The field dependence of thermal conductivity is measured as a change of the tem-

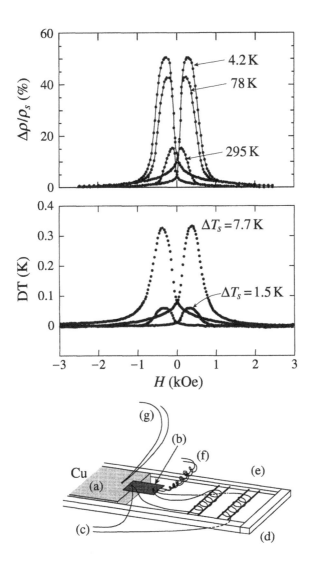

Figure 1.13: (Upper) Field dependence of MR ratio at 295 K, 78 K and 4.2 K. (Middle) Thermal conductivity vs magnetic field for [NiFe 20 Å/Cu 40 Å/Co 20 Å/Cu 40 Å]×50. DT means the temperature change caused by the field sweeping, which corresponds to the thermal conductivity. $\Delta T$ is the initial temperature difference at 3 kOe. (Lower) Schematic illustration of the structure of the sample holder. (a) Cu heat sink, (b) sample, (c) Au(Fe)-chromel differential thermocouple, (d) bakelite flame, (e) nylon threads, (f) $RuO_2$ heater, and (g) thermocouple [34].

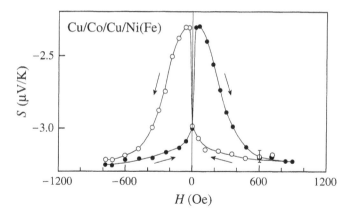

Figure 1.14: Thermopower at 80 K of Cu/Co/Cu/NiFe as a function of magnetic field [35].

perature difference. If the conductivity is enhanced, the temperature difference should decrease. To the contrary, if it is reduced, the temperature difference should increase. The difference of temperature was measured with changing the external field and the result is shown in Fig. 1.13. Two curves show the results with the initial temperature differences, 7.7 K and 1.5 K, respectively. The profile of obtained curves are very similar to the MR curves shown in the same figure. At almost the same magnetic field, both electric resistivity and thermal conductivity curves exhibit pronounced peaks. At this field, the electric resistivity becomes the maximum because the magnetic configuration is turned out to be antiparallel and at this situation, the thermal conductivity becomes the poorest. The reason is attributed to the contribution of conduction electrons to the thermal conductivity whose mean free path has been varied by the GMR effect. The change of the temperature difference is much more sensitively observed in the curve with the initial temperature difference of 7.7 K than in that of 1.5 K. The sample temperature is approximately 10 K for both cases. The quantitative estimation of thermal conductivity change in the sample with subtracting the gross contribution from the thick glass substrate cannot be accurate but the change of thermal conductance in the sample due to the magnetic origin was roughly estimated to be 45%. Here it is evidenced that a novel variation of the GMR effect in multilayers is the magnetic field-dependent thermal conductivity. A sizable field dependence of thermal conductivity is observed also at higher temperatures, 80 and 300 K, but a potential for technical applications is not clear. The change of conductivity caused by the magnetic field appears to be too small to use for any technical purposes.

Thermopower of the same non-coupled type multilayer was measured by Sakurai et al. [35]. The field dependence in Fig. 1.14 clearly shows the difference of the thermopower between spin parallel and antiparallel states. The profiles of thermopower vs magnetic field curve is consistent with those of the electric resistivity and there it is no doubt that the observed peaks are caused by the magnetic structure of antiparallel alignment. The sign of thermopower is negative and the absolute value of antiparallel state is smaller than that of parallel. The sign of thermopower was found to be positive for Fe/Cr multilayers while negative for Cu/Co multilayers [36, 37] and the difference of the sign was argued in terms of the electronic band structures for Co/Cu and Fe/Cr multilayers by Tsymbal *et al.* [38].

## 1.2.5    Non-coupled sandwiches (Spin valve systems)

Sandwich systems consisting of magnetic/non-magnetic/magnetic trilayers are important particularly from the two viewpoints. One is to study the interlayer coupling as was mentioned already. The other is for technical applications. Almost at the same time with the report on non-coupled type GMR in multilayers [26], Dieny *et al.* published a paper on non-coupled GMR in trilayer systems and named "spin-valve" [29]. They have differentiated the coercive force of one magnetic layer by coupling with an antiferromagnetic FeMn layer. The initial structure of a spin-valve sample is: [NiFe 150 Å/Cu 26 Å/NiFe 150 Å/FeMn 100 Å]. Namely two magnetic layers of NiFe are separated by a Cu layer and an FeMn alloy layer was attached to one of the NiFe layer as a magnetic anchor. FeMn is an antiferromagnet with the Néel temperature of about 150 °C. Because of the exchange anisotropy between FeMn and NiFe, the effective coercive force of the NiFe layer is enhanced up to more than 100 Oe. Since the other NiFe layer behaves as very soft, two magnetizations are oriented to be antiparallel with each other in a certain region of external field. Because of the GMR effect, the resistivity increases if the two magnetizations are antiparallel and decreases if parallel. Owing to the easiness in controlling the coercive force and also the resistance is of an appropriate size, the spin valve sandwich system has been recognized to be suitable for application and actually used as commercial GMR heads.

     In the case of spin valve systems, antiferromagnetic layers are required to be rather thick and it is hard to repeat the deposition to make multilayers. Therefore, spin valve samples are in principle sandwiches practically including only two magnetic layers. Initially, it was thought to be inevitable that the MR ratio of a sandwich is smaller than that of a multilayer. However, recently there have been many reports that the MR ratio can be significantly enhanced in spin valve systems with sophisticated structures. In a sandwich structure, conduction electrons meet surfaces many times, and scatterings by non-magnetic origins at surfaces and interfaces will reduce the effective MR ratio. If surfaces (and interfaces) act as mirrors and electrons are reflected

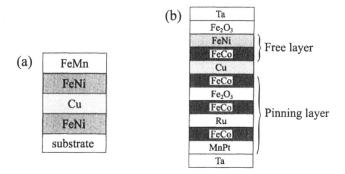

Figure 1.15: Schematic illustration of spin-valve structures. (a) Initial, and (b) Sophisticated.

specularly, there is no increase of the background resistivity and thus no loss of the MR ratio. Egelhoff *et al.* [39] and other recent papers [40] reported that using specularly reflecting layers, the MR ratio is kept to be fairly large; 20% at room temperature. The specular reflectivity greatly depends on the material and the quality of the layer and for example an oxide layer (e.g., $Fe_2O_3$) are used as specular layers. It was also reported that an oxide layer with the thickness of a few Å inserted in a magnetic layer works as a reflective layer and subsequently enhance the MR ratio [41]. In order to enhance the coercivity of pinning layer, $Co/Ru/Co/Fe_2O_3$ structure is very effective. The interlayer antiferromagnetic coupling between Co layers across a Ru layer is known to be very strong and the cancellation of magnetization using the $Co/Ru/Co$ structure can increase greatly the coercivity of the pinning layer. In this structure, an $Fe_2O_3$ layer is expected to behave as a specular reflective layer.

The spin dependent scattering probability is increased if the surface magnetic moments of the magnetic layers are larger. Initially, magnetic layers are often composed by NiFe but in recent samples surfaces of NiFe layers are modified by depositing thin FeCo layers to enhance the surface magnetic moments. It becomes also common to use FeCo layers instead of NiFe layers. Although the spin valve structure in principle consists simply of two magnetic layers, such a sophisticated design as shown in Fig. 1.15 is constructed to use for commercial read-out heads responsible to a recording density of more than $10\,Gbit/in^2$.

## 1.2.6    Extension of studies on GMR-related phenomena

From GMR, we have learned that the resistivity depends on the magnetic structure and an unexpectedly large MR effect is able to be induced by magnetic structure transformation. Not only in metallic multilayers, but also in some other systems, similar phenomena to GMR have been found. Two groups in USA reported that GMR phenomena take place in granular systems where ultrafine ferromagnetic particles are dispersed [42, 43]. A typical example is Cu/Co system, in which precipitated Co particles are dispersed in a Cu matrix. In case that the concentration of Co is 5%, for instance, the average particle size of Co is about 30 Å in diameter. In the absence of external field, individual magnetizations of Co particles are oriented at random and those are saturated by applying an external field. The electric resistance shows a decrease associated with the change of magnetic structure. Namely, the transformation of magnetic structure from random to parallel in granular systems corresponds to that from antiparallel to parallel in multilayers and then the spin-dependent scattering probability has eventually decreased. At least for the MR behaviors, a randomly oriented magnetic structure is considered to be analogous to an antiparallel alignment. It is not certified that the magnetizations are ideally oriented at random in granular systems. If any coupling or dipole interaction exists between each particle, there may be a tendency for adjacent particles to be oriented antiparallel. The overall magnetic structure in granular system without external field is approximated as random but the relative angle between magnetizations in adjacent particles may be closer to antiparallel than random. A characteristic of GMR in granular systems is the field dependence; a remarkable large external field is required for saturation (Fig. 1.16) [44]. The magnetic hardness of the granular systems may be an evidence that the individual particles are not magnetically free but are coupled with each other. In a rather high field region where the magnetization has been almost saturated, the resistance still shows a definite decrease with increase of the field. It is suggested that although the majority of magnetization is saturated, some minor fractions are still keeping non-collinear alignments, which are very influential to the conductivity. If the magnetic spins on the surface of precipitated particles have some canting, the field dependence of resistance may be accounted for. This result suggests that the spin-dependent scattering sensitively depends on the spin structure of very specific parts, such as interface sites. In general, granular systems are not magnetically soft and the saturation of resistivity requires a larger field than the saturation field for bulk magnetization. From the practical reason that the resistivity at saturation is hard to estimate, MR ratio in granular system is often expressed as a relative decrease from the zero field value. If the system is magnetically hard, it is common to express the MR ratio relative to zero field resistivity, similar to the case of granular systems. The resistivity of granular systems varies with increase of external field up to very high fields but is not sensitive at low fields. Because of this feature, the potential of granular systems for technical applications seems to be limited. Even when the matrix is a simple non-magnetic noble metal like Cu, the interaction between particles is not clearly understood. If it is magnetic, the situation is more complicated. Somsen *et al.* reported the GMR

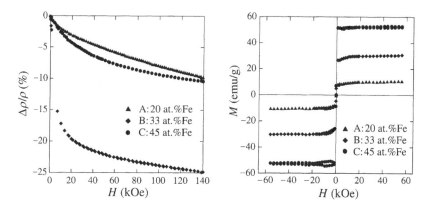

Figure 1.16: Magnetization and resistivity at 4.2 K of a granular system (Fe/Ag films) under very high fields [44].

behaviors on precipitated Fe particles in Cr matrices [45]. They add V and Mn to vary intentionally the magnetic transition temperature of Cr. By measuring the temperature dependence of the MR ratio, they found that the MR ratio becomes large in the vicinity of the Cr-matrix Néel temperature.

Studies on GMR effect using tunneling currents (TMR) have been greatly developed recently [46, 47]. Although pioneering experiments have been initiated even before the discovery of GMR in metallic multilayers, the MR ratio was small and reproducibility was rather poor. As is described in [46, 47], updated studies show that TMR in the systems consisting of ferromagnetic metals separated by insulating layer exhibit fairly large MR ratios at room temperature with good reproducibility. Similarly to the spin valve systems it is possible to design a magnetically soft TMR system with a sandwich structure. As for technical applications, TMR systems are regarded to have high potentials [48]. TMR in a granular system has been reported by Mitani *et al.* [49]. The samples consisting of metallic magnetic particles isolated by Al oxide layers are prepared by a codeposition of $3d$ metal, as a result of natural oxidation. The granular TMR systems are not magnetically soft similarly to the granular GMR systems but from fundamental viewpoints those behaviors are of remarkable interests, as are mentioned in Sec. 4.3.4.

The GMR-like phenomena are observed also in some compounds. If the magnetic structure is modified by external field, the resistivity may be varied owing to the change of spin-dependent scattering probability. LaNiGa shows a metamagnetic transition and spin alignment changes from antiparallel to parallel. Then the resistivity drops due to the decrease of spin-dependent scattering [50]. The behavior is very similar to the GMR effect in multilayer system. There have been known rather many compounds which exhibit metamagnetism, in which a ferromagnetic alignment is induced by external fields.

That is a change of magnetic structure from antiparallel to parallel. However, there are few resistivity measurements from a viewpoint of MR. It might be possible to find a large MR effect in a metamagnetic compound where the resistivity may change associated with the magnetic structure transformation due to the spin-dependent scattering. The external field required for meta-magnetic transition is normally not small but it is important to re-examine metamagnetic transitions for the basic understanding of MR effect. Perovskite compounds of $LaSrMnO_3$ type show a very big resistivity change associated with a magnetic transition and the tremendously large MR ratio has been named colossal magnetoresistance, CMR [51]. In the case, the magnetic transition is not a simple reorientation of spins but the electronic structure seems to have drastically changed. Reviews on CMR studies are given in another book of the same monograph series [52].

## 1.2.7 Geometry of GMR

Usually resistivity measurements for thin film specimens with metallic conductivity are carried out in a conventional geometry that the electric current direction is in the film plane, which is called CIP geometry (current inplane). For metallic films, resistivity measurements in CIP geometry are very simple. In contrast, to measure the resistivity along the normal direction to the plane, which is called CPP geometry (current perpendicular to plane), is very inconvenient because first of all the resistance of metallic films in CPP geometry is extremely small. Before the discovery of GMR, it had been naively considered that metallic multilayers might exhibit some striking characteristics only when a transport measurement is made in CPP geometry but fortunately the primitive concept was not correct. It has been proved that the GMR effect is remarkably large in the measurements in CIP geometry and many interesting features have been found in the magnetotransport properties of metallic multilayers in CIP geometry. Already industrial applications are realized using GMR phenomena in CIP geometry.

However, the naive concept that the GMR effect can be larger in CPP geometry than in CIP is certainly correct. There are several theoretical studies on CPP-GMR suggesting a significantly large enhancement of MR effect due to the CPP geometry [53, 54]. More generally, it is no doubt that the study of CPP transport phenomena of metallic multilayers is essentially important because the direction of artificially designed structure which we can construct is along the film normal. It is possible to control each layer thickness on an atomic scale and even monolayer/monolayer superstructures are realized in certain combinations of metals [55]. Such novel structures must be reflected in the transport properties in CPP geometry. Besides the enhancement of GMR effect, MR measurements in CPP geometry have a definite advantage for the fundamental study of the mechanism of GMR. Namely, the electric

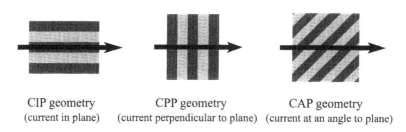

CIP geometry      CPP geometry      CAP geometry
(current in plane)    (current perpendicular to plane)   (current at an angle to plane)

Figure 1.17: CIP, CPP, and CAP geometries.

current in the normal direction to the film is regarded to be constant in the sample and therefore the simple series resistor model is able to be applied, in which the total resistance is assumed to be the sum of each component. On the other hand, in CIP geometry, the electric current density depends on the resistance of each layer, which is not homogeneous even in an individual layer of a multilayer structure. An accurate estimation of real electric current density in CIP geometry is almost impossible and a quantitative analysis is applicable only for CPP-MR data. Thus the significance of GMR measurements in CPP geometry is clear and already a review paper has been published [56]. In Fig. 1.17, the concepts for CPP and CIP geometries are illustrated together with CAP, which is an intermediate case between CIP and CPP as described later.

Although CPP-GMR measurements are very attractive, because of experimental difficulty, there have been only a few pioneering experiments on metallic multilayers. Pratt Jr. *et al.* have challenged for the first time to measure the very small resistivity in CPP geometry [57]. If the total thickness of a multilayer is $1\,\mu$m, the resistance of a sample with $1 \times 1$ mm area is expected to be $10^{-7} \sim 10^{-8}\,\Omega$. A typical current of $10\,$mA yields $10^{-9} \sim 10^{-10}\,$V. For the measurement of very small voltage, they have prepared multilayer samples sandwiched in superconducting Nb lead layers and measured by using SQUID technique. The sample structure is schematically shown in Fig. 1.18(a). In order to hold the superconductivity of the lead layers, the measuring temperature was inevitably in the liquid helium region and the external field is also limited. In their initial report, the CPP-GMR ratios of Ag/Co multilayers were compared with the CIP-GMR. In such a coupled type GMR multilayers, the spacer layer thickness is required to be specific values to realize an antiferromagnetic interlayer coupling. Instead of an antiferromagnetically aligned structure, they have compared the resistivity of a ferromagnetic state with uncorrelated state initially formed in the "as-prepared" sample before applying magnetic fields. Using this method, they compared the CIP-GMR and CPP-GMR in the same Co/Ag multilayers and observed that the CPP-GMR

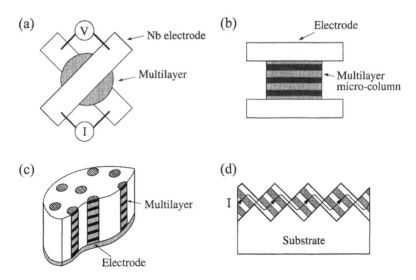

Figure 1.18: Various sample structures for the CPP-MR studies.

ratio is by 3 to 10 times larger than the CIP-GMR ratio. The possibility to enhance the MR ratio in the CPP geometry has been clearly indicated by the experiment. The working temperature, however, is limited to be low as far as superconducting leads are used and therefore for instance the temperature dependence of MR effect is not able to study.

The microfabrication technique has been introduced by Gijs *et al.* to the GMR studies in CPP geometry [58]. They have prepared Fe/Cr and Co/Cu multilayer samples with micropillar structures by conventional lithography and etching techniques and studied the GMR effect in CPP geometry. The cross section of the pillar shaped sample is also schematically shown in Fig. 1.18(b). In their experiments, the thickness (pillar height) is about $1\,\mu$m, while the area is more than $6\,\mu$m$^2$. The aspect ratio is much lower than 1. The typical resistance value is in the m$\Omega$ region and technically the measurement is not very difficult. Although the height is not sufficiently larger than the width, and therefore the current direction is not regarded to be ideally CPP, the result shows apparently the characteristic of CPP that the MR ratio is considerably larger than that of CIP in the simultaneously prepared sample. In contrast to the previous experiments using superconducting electrodes, it is of course possible to measure the temperature dependence of CPP-GMR as shown in Fig. 1.19. The real current density distribution however is a complicated problem, as revealed in the argument by Gijs *et al.* [59].

There has been only a few studies since then in this direction; namely to prepare artificial pillar structures of GMR multilayers using microfabrication

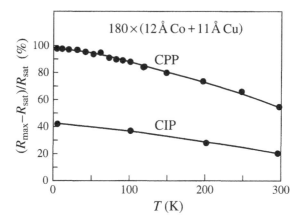

Figure 1.19: Temperature dependence of the CPP-MR of Co/Cu pillar structure [58].

techniques. The reason seems to be simply the inconvenience of mircofabrication techniques for metallic systems, which require expensive equipments and some know-hows [60]. Compared with semiconductors, the application of microfabrication to metallic magnetic materials is not yet routine. Recently, microfabrication facilities become popular for magnetic materials. GMR in micropillars connected in series have been measured recently [61].

Another approach to CPP-GMR was made by using microwires of multilayers prepared in pores of membrane filters of polycarbonate by means of electrodeposition [62]. As illustrated in Fig. 1.18(c), pores with the diameter of $30 \sim 300$ nm are constructed in membranes by irradiation techniques and multilayers are synthesized in each pore. Before the multilayer deposition, a metal layer of Cu or Au is deposited on one side of the membrane, which acts as a cathode in the electrodeposition process. Then, the shape of the obtained samples is an ideal wire and concerning the aspect ratio, the samples are very suitable for CPP-GMR studies. The current direction is approximated to be in the wire axis and the sample resistance is much larger than the contact resistances. Using samples fabricated by this method, an enhancement of the GMR effect has been reported but the MR ratio is not remarkably large yet. A problem in this method seems to be the quality of multilayer structures. Using electrodeposition it is generally difficult to control the each layer thickness accurately and also to keep the sharpness of interfaces. However, the shape of prepared nanowire is indeed excellent. Of course it is possible to remove the wire samples from the membrane matrix. There must be considerable potentials for technical applications.

Preparation of V-groove structure on the surface of substrate is an established technique in the semiconductor field. The typical procedure is shown in

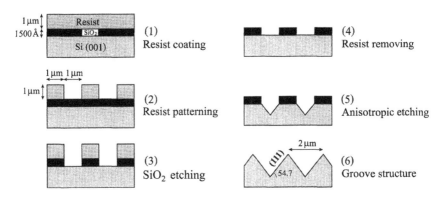

Figure 1.20: Procedure for preparing a substrate having a surface with V-groove microstructure.

Fig. 1.20. By photo- or electron-beam lithography, a stripe pattern with $0.5\,\mu m$ pitch for instance is fabricated on a surface of Si(100) covered by a thin $SiO_2$ layer. By etching, $SiO_2$ stripes are prepared. Then the Si is etched by using HF solution. Because the etching for a (100) surface is 2 order faster than for the most stable (111) surface, finally (111) planes appear on the surface and as a consequence V-shaped groove is formed. The angle of the V-groove peak is therefore always 54.7°, which is the intercept of two (111) planes. Using multilayers prepared on V-groove substrates, several novel experiments have been carried out in relation with GMR studies. Concerning the geometry issue, multilayers prepared on V-groove substrates have been utilized to measure the GMR effect in CAP and pseudo-CPP geometries.

Ono *et al.* prepared non-coupled GMR multilayers, Cu/Co/Cu/NiFe, deposited on a substrate with V-shaped grooves as sketched in Fig. 1.21(a) [63]. If the deposited film is thick enough, then the electric current in the sample is nearly parallel to the original plane of substrate. If the conductivities of the constituents are different, the current direction is not exactly straight but in metallic systems, it may be approximated to be a straight line. Then, the current direction has a certain angle to the interfaces in the multilayered structure, that is 54.7°. This is an intermediate geometry between CPP and CIP, and defined as CAP (current at an angle to plane). The MR ratio of the same sample is measured by taking the current direction perpendicular to the grooves. In this direction, the current is parallel to all the interfaces and the CIP-MR are measured. The observed MR ratio of CAP geometry is definitely larger than that of CIP (Fig. 1.21(b)). If the current direction is rotated in the original plane of the substrate, the angle between current and interfaces is varied from 0 to 54.7°. The MR ratio in this angle region is able to approximate with a simple cosine law as shown in Fig. 1.21(c), and therefore

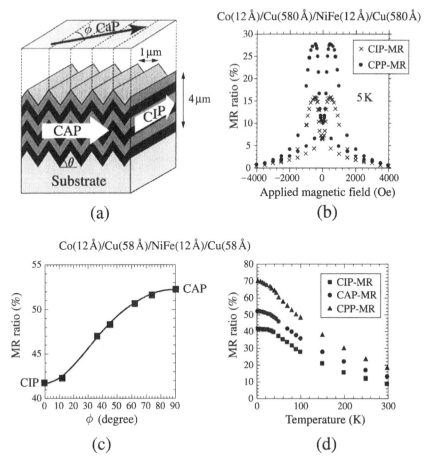

Figure 1.21: (a) Schematic illustration of the multilayer sample structure for CAP-MR studies. (b) MR ratios at 5 K for the same samples in CAP and CIP geometries. (c) CAP-MR ratio as a function of the angle $\phi$. The definition of $\phi$ is given in Fig. 1.21(a). (d) Temperature dependence of CIP, CAP and CPP-MR [63].

it seems possible to extend the curve up to 90°, i.e., the CPP-geometry. The result shown in Fig. 1.21(d) is the temperature dependence of observed CIP- and CAP-MR, together with the extrapolated CPP-MR values.

Gijs *et al.* deposited multilayers from a tilted direction, perpendicularly to one of the side walls of V-groove structure [64]. If the current direction is as shown in Fig. 1.18(d), an MR ratio of CPP geometry should be measured. The current direction cannot be exactly perpendicular to the multilayer interfaces and the distributions of direction and density of current are not trivial. The

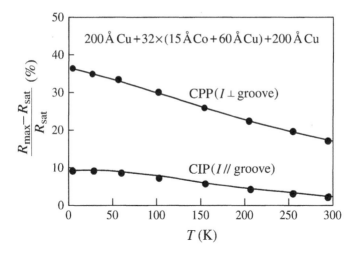

Figure 1.22: Temperature dependence of pseudo CPP-MR and CIP-MR ratios for a multilayer [Co 15 Å/Cu 60 Å] × 32. (Refer Fig. 1.18(d) for the sample structure.) [64]

real geometry is therefore a "pseudo-CPP". The average current direction is however close to perpendicular and the observed MR ratio is significantly large, as shown in Fig. 1.22, in comparison with the CIP-MR in the same sample. Instead of microfabricated V-groove substrates, Encinas *et al.* have used step-bunched vicinal Si(111) substrates [65]. They found that the MR ratio of a Co/Cu/NiFe trilayer on a vicinal surface with current perpendicular to the steps has increased by 70% compared with that with current parallel. In these pioneering works, it has been experimentally recognized that the MR ratio is certainly enhanced by the effect of geometry. However, from technical reasons, studies on CPP-GMR is not yet popular. For fundamental research on the mechanism of CPP-MR, non-coupled multilayer systems are advantageous to construct antiparallel alignment of magnetizations artificially. In the case of coupled systems, it is laborious to control the thickness to be the optimum value for antiferromagnetic interlayer coupling.

Concerning the materials to realize the GMR effect, material hunting has been made to a considerable extent for CIP-geometry. In contrast, in CPP geometry the material hunting is not sufficient yet. The freedom in the material selection in CPP is broader than in CIP. For instance, if the conductivity of non-magnetic spacer layer is insulating, the current will flow only in magnetic metallic layers and eventually no GMR effect appears. In CPP geometry, on the other hand, the spacer layer can be semiconductive or even insulating. If the spacer layer is an insulator, tunneling GMR is expectable. Tunneling MR only occurs in CPP geometry and thus TMR is a kind of CPP-MR. In several experiments, metallic GMR system is combined with semiconductors such as Si, and GMR of hot electrons is observed [66, 67]. The reason of large MR ratios observed in those experiments should be partially the CPP geometry.

## 1.3    Studies on nanomagnetic systems using GMR effect

Several experiments on magnetic nanowires are described in this section. In an ultrathin wire of a ferromagnetic material, the direction of magnetizations limited to be parallel to the wire axis because of the shape anisotropy. As illustrated in Fig. 1.23, if there exists only one domain wall in a wire, the magnetization reversal of the wire takes place associating with the propagation of the domain wall. The concept of domain wall has been established long time ago but the dynamical behaviors of domain walls are still a crucial issue from technical aspects. From fundamental viewpoints, magnetic domain walls in mesoscopic scales have attracted renewed attention since the possibility of macroscopic quantum tunneling (MQT) process was predicted. Pioneering studies on MQT in mesoscopic systems have been reported by several groups [68]. If the wire cross-section is small, the domain wall size should be small and then the domain wall is expected to behave as a quasiparticle. Magnetization reversal phenomena in a nanoscale ferromagnetic wire are therefore of particular interest but to observe a magnetization change in a single nanowire sample is a hard experiment. Measurements of resistivity in wire samples are rather easy but information on magnetism is normally not furnished. A domain wall means a region with a non-collinear spin arrangement and it should contribute to the resistivity. There have been several experiments to make clear a resistance due to the disordered spin structure but even concerning the sign of the resistance the results are not consistent. Some insist a decrease of resistance due to domain wall and others an increase [69]. In any case, a contribution of domain wall to the resistivity is fairly small and is a minor effect compared to the AMR effect. The resistance will be invariant during the magnetization reversal as shown in Fig. 1.23 and therefore resistivity measurements seem to be useless for the study of magnetization reversal process in nanowires.

By utilizing GMR effect, however, resistivity measurements become a sensitive tool to observe magnetization reversal phenomena. If we prepare a wire sample consisting of trilayer exhibiting non-coupled type GMR, the resistivity shows the ratio of antiparallel alignment of magnetizations. As illustrated in Fig. 1.23(d), if the magnetization is reversed by the propagation of a single domain wall, the degree of reversal is observed as a change of resistance. An example of experimental results on a multiwire sample is shown in Fig. 1.24. Experiments using substrates with V-shaped grooves for the study of GMR in CAP geometry has been mentioned already. V-groove substrates are also

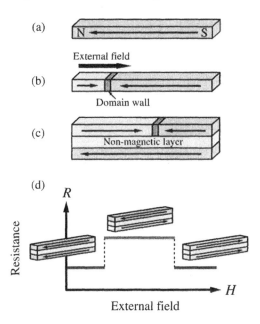

Figure 1.23: (a), (b) Magnetization reversal of a wire associating with the propagation of the domain wall. (c) Structure of trilayer system. (d) Resistance change associating with magnetization reversal.

useful to prepare an array of nanowires [70]. If the deposition is made in a tilted angle, the deposited film is divided into a wire shape at each groove and subsequently an array of wires is prepared. Using a V-groove with a pitch of $0.5\,\mu$m, wires with $0.3\,\mu$m width are obtained. In a sample on a Si substrate with an area of 10mm × 10 mm, about 20 000 wire pieces are included. Each wire consists of five layers with the nominal structure of NiFe 100 Å/Cu 100 Å/NiFe 10 Å/Cu 100 Å/NiFe 100 Å. It has been reported that coercive force of NiFe nanowire increases with increase of thickness in the region up to about 300 Å [71] and NiFe nanowires with two thicknesses change the direction of magnetization at different switching fields. The coercive force of NiFe 10 Å layer is fairly small but that of 100 Å layer is larger than 100 Oe. Therefore, the samples including NiFe layers with two thicknesses exhibit non-coupled GMR phenomena, similar to the spin-valve system. Figure 1.24 shows the magnetization and resistivity of the wire array sample. In the magnetization curve, a small lump is observed at about 20 Oe corresponding to the reversal in 10 Å layer. The change is so small that any quantitative analysis is hard to apply. On the other hand, the resistivity change at this field is strikingly large, indicating apparently the magnetizations being antiparallel in the field region from 20 to 100 Oe. It is noteworthy that owing to the GMR effect, the resis-

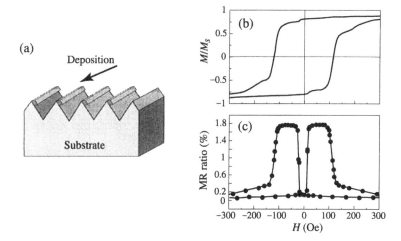

Figure 1.24: (a) Structure of wire array deposited on a V-groove substrate. The arrow indicates the direction of the deposition. (b) Magnetization and (c) resistivity at 300 K of wire arrays with a nominal structure of [NiFe 100 Å/Cu 100 Å/NiFe 10 Å/Cu 100 Å/NiFe 100 Å] as a function of external field [71].

tivity change is much more remarkable than the corresponding change in the magnetization curve. This result proves that the resistivity measurements utilizing non-coupled GMR effect are useful to study the magnetization reversal phenomena in wire systems. Concerning magnetization reversal phenomena, from resistivity measurements on multiwire samples, we obtain a statistical information and can compare with magnetization change to some extent. For a single wire sample, magnetization measurements are not able to be applied but resistivity measurements are applicable as well as multiwire samples.

Samples for the study of magnetization reversal having such a pattern as shown in Fig. 1.25(a) were prepared by using electron-beam lithography apparatus with a lift-off method [72]. The content is the following trilayer structure, NiFe 200 Å/Cu 100 Å/NiFe 50 Å, and the wire width is 0.5 $\mu$m. The distance between the probes for resistivity measurements is 20 $\mu$m. In addition, the sample has a neck point at 1/3 distance from one voltage probe, where the width was artificially narrowed (nominally 0.35 $\mu$m). Domain wall propagation is expected to be pinned at the neck. The resistivity as a function of the applied field at 300 K is shown in Fig. 1.25(c). Prior to the field sweeping, a magnetic field of 100 Oe was applied in order to align all magnetization parallel. In the figure, an enhancement of resistivity owing to the GMR effect observed in the course of field sweeping is reproduced. Both the increase and the decrease of resistivity have two steps and the jumps between the steps are very sharp. The relative height ratio of the two steps is 1:2 at both cases, which means that

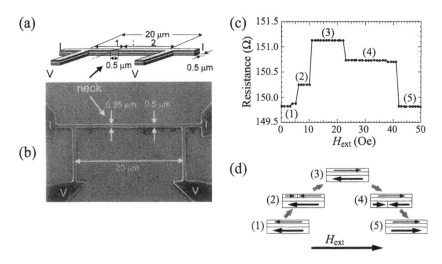

Figure 1.25: (a) Schematic illustration of the sample and (b) SEM image. The sample consists of a NiFe(200 Å)/Cu(100 Å)/NiFe(50 Å) trilayer. (c) Resistance as a function of the external magnetic field at 300 K determined by the four-point dc technique. (d) Schematic diagram of the magnetic domain structures [72].

1/3 of the magnetization is reversed first and after a while remaining 2/3 is reversed. As expected, a domain wall is trapped at the neck point and when the external field exceeds a certain critical value, the domain wall is depinned or another domain wall comes from the other side to complete the magnetization reversal. For this single wire sample, magnetization measurements are not applicable but from the resistivity measurements the process of magnetization reversal is clearly checked. It is no doubt that the magnetic structure at each stage is as illustrated in the figure. The result indicates that the resistivity is kept constant at each stage but the resistivity change at the transition between stages is very sudden. The process of magnetization reversal is therefore regarded as a propagation of single domain wall and the velocity should be very high.

The velocity of domain wall propagation is in principle able to estimate from the time dependence of resistivity. However, if the velocity is very fast and the probing length is too short, the variation of resistivity as a function of time is not easy to measure. For the velocity measurements, a sample with 100 times longer distance (2 mm) was prepared with the structure of NiFe 400 Å/Cu 200 Å/NiFe 50 Å and 0.5 $\mu$m width, having no neck point [73]. The resistivity as a function of external field at 77 K is shown in Fig. 1.26(a). After applying 500 Oe in the negative direction, the resistivity was monitored with sweeping

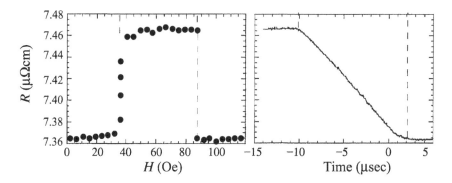

Figure 1.26: (a) Resistance as a function of the external magnetic field at 77 K. (b) Time variation of the resistance during the magnetization reversal of the 400 Å NiFe layer [73].

the field in the rate of 20 Oe/s. The resistivity enhancement due to the GMR effect is found in the region of 35 and 85 Oe and the increase and the decrease correspond to the reversal in 50 Å and 400 Å NiFe layers, respectively. The time dependence was measured for the sharp switching in the thicker layer by using a digital oscilloscope. As shown in Fig. 1.26(b), the change of resistivity has taken place in the time interval of 11 $\mu$sec. Because the field sweeping rate is very slow, the external field during the time dependence measurement is actually constant, i.e., 88 Oe. The result indicates that the resistivity change is almost linear to the time variation, which suggests that the magnetization reversal is caused by a single domain wall moving with a constant velocity. Since the time length to travel in the distance of 2 mm is measured to be 11 $\mu$sec, the velocity is calculated to be 182 m/s. The domain wall in a wire is regarded as quasiparticle moving in a very fast but almost constant velocity. The experiments introduced above prove that behaviors of a single domain wall are able to be monitored by the resistivity measurements utilizing the GMR principle. This method is useful for the study of magnetization reversal phenomena in mesoscopic systems and dynamical behaviors of domain walls.

In the initial stage of the experiment, the whole body of the wire sample including the terminal parts has been composed of the same magnetic structure for the convenience of the sample preparation and accordingly the observed switching field values are considerably distributed. Possibly because of neck or elbow parts included in the sample, the structure of domain wall produced at each run is not the same from a microscopic point of view. Improvement of the quality of wire samples has been made by preparing the terminal parts by a non-magnetic metal (Cu) and only the wire part was constructed by magnetic layers [74]. In addition, as illustrated in Fig. 1.27(b), a pad with a wide area was attached at one end of the wire to specify the nucleation site and the

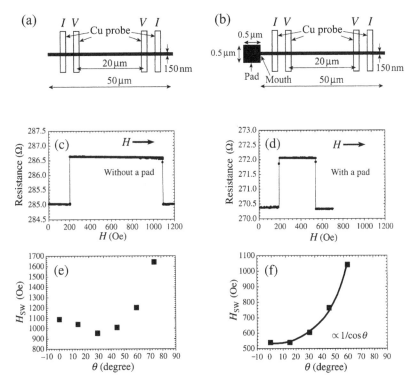

Figure 1.27: (a)(b) Schematic illustrations of the top views of the two types of wires with and without a pad. (c)(d) Resistance as a function of the external magnetic field $H$ parallel to the wire axis at 300 K. (e)(f) Angular dependence of the switching field $H_{SW}$ of the Co layer at 300 K. The magnetic field was applied in the plane of the substrate at an angle $\theta$ with respect to the wire axis [74].

direction of domain wall propagation. The results in Fig. 1.27 were obtained for a sample whose magnetic layers are consisting of permalloy and cobalt. Similarly to the results in Fig. 1.25, abrupt changes of resistivity due to the individual reversals of magnetization in permalloy and cobalt are observed. For this sample, the values of external field for the reversal are reproducibly obtained. The reversal process is considered to be the following: The nucleation starts in the pad part under a rather low field but a domain wall is trapped at the mouth and does not proceed into the wire part until the external field exceeds a certain critical field. This model is verified by the measurements with changing the direction of external field relative to the wire axis. With increase of the angle $\theta$, from the axis, the switching field increases as shown in

Fig. 1.27. However, for the sample with a pad, $H \cos \theta$ is found to be constant. Namely, the component of magnetic field to the direction of the axis is the same. The domain wall starts to propagate when the magnetic field strength to the propagation direction reaches to a critical value. Each resistivity measurement as a function of the angle shows very sharp changes suggesting all the process is due to a single domain wall propagation. The curvature of resistivity at higher angles is attributed to the AMR effect and also the deviation of the magnetization alignment from antiparallel caused by applying an external field at an angle to the wire axis. The domain wall formation and movement are thus satisfactorily controlled and this methodology seems to be promising to study the magnetization reversal phenomena in samples with smaller sizes.

## 1.4   Summary

Experimental studies on GMR and related phenomena have been surveyed briefly. The discovery of GMR revealed us that the influence of magnetism, or spin polarization, on transport properties is significant. Investigations are being made extensively for various types of MR phenomena with hoping to explore novel phenomena caused as an interplay between magnetism and transport. An important key word for further progress seems to be micro- or nanofabrication. A newer MR effect will appear on the stage where the sample structure is controlled on an atomic scale. It is less than 10 years from the discovery of GMR to the commercial application as the GMR head, which is surprisingly short. This fact evidences that the field of magnetism studies has a very high activity and in this field the distance between basic physics and application is rather small. At the moment, the spin-valve system is used as the GMR head and the specific properties satisfy to cover the density of recording up to several tens Mbit/in$^2$. However, some years later, read out sensors with higher sensitivity will be required. In future also, the studies on MR properties should be of great importance both from basic and technical viewpoints. In this article, only a limited number of publications were introduced. This field is rapidly expanding and it is practically impossible to cover all the publications in a small volume. The author apologizes that many updated and distinguished works could not be cited here.

The author thanks to the collaborators, whose names are shown as the co-authors of the publications in the reference list. The works of the author were supported by Grants-in-Aid from Monbusho and NEDO.

# References

[1] M. N. Baibich, J. M. Broto, A. Fert, F. Nguyen Van Dau, P. Etienne, G. Creuzet, A. Friederich and J. Chazelas, Phys. Rev. Lett. **61**, 2472 (1988).

[2] M. Julliere, Phys. Lett. **54**, 225 (1975).

[3] S. Maekawa and U. Gäfvert, IEEE Trans. Mag. **18**, 707 (1982).

[4] E. Velu, C. Dupas, D. Renard, J. P. Renard and J. Seiden, Phys. Rev. B **37**, 668 (1988).

[5] P. Grünberg, R. Schreiber, Y. Pang, M. B. Brodsky and H. Sowers, Phys. Rev. Lett. **57**, 2442 (1986); F. Saurenbach, U. Walz, L. Hinchey, P. Grünberg and W. Zinn, J. Appl. Phys. **62**, 3473 (1988).

[6] G. Binasch, P. Grünberg, F. Saurenbach and W. Zinn, Phys. Rev. B **39**, 4828 (1989); M. Vohl, J. Barnas and P. Grünberg, Phys. Rev. B **36**, 12003 (1989).

[7] C. Carbone and S. F. Alvarado, Phys. Rev. B **39**, 2433 (1987); S. F. Alvarado and C. Carbone, Physica B **149**, 43 (1988).

[8] For magnetic multilayer studies before GMR, see references cited in *Metallic Superlattices*, edited by T. Shinjo and T. Takada (Elsevier Sci. Pub., 1987).

[9] A. Cebollada, J. L. Martinez, J. M. Gallego, J. J. de Miguel, R. Miranda, S. Ferre, F. Batallan, G. Fillion and J. P. Rebouillat, Phys. Rev. B **39**, 9726 (1989).

[10] N. Hosoito, S. Araki, K. Mibu and T. Shinjo, J. Phys. Soc. Jpn. **59**, 1925 (1990).

[11] R. Schad, C. D. Potter, P. Belien, G. Verbanck, J. Dekoster, G. Langouche, V. V. Moshchalkov and Y. Bruynseraede, J. Magn. Magn. Mater. **148**, 331 (1995).

[12] A review paper for theories on GMR for instance is: P. M. Levy, Solid State Phys. **47**, 367 (1994). Experimental results on GMR are comprehensively listed in the following, A. Barthélémy, A. Fert and F. Petroff, *Handbook of Magnetic Materials*, Vol. 12, edited by K. H. J. Bushow (Elsevier, 1999), p. 1.

[13] F. J. Himpsel, J. Phys. Condens. Mater. **11**, 9483 (1999).

[14] D. H. Mosca, F. Petroff, A. Fert, P. A. Schroeder, W. P. Pratt Jr. and R. Loloee, J. Magn. Magn. Mater. **94**, 1 (1991).

[15] S. S. P. Parkin, R. Bhadra and K. P. Roche, Phys. Rev. Lett. **66**, 2152 (1991).

[16] S. S. P. Parkin, N. More and K. P. Roche, Phys. Rev. Lett. **64**, 2304 (1990).

[17] S. S. P. Parkin, Phys. Rev. Lett. **67**, 3598 (1991).

[18] J. Unguris, R. J. Celotta and D. T. Pierce, Phys. Rev. Lett. **67**, 140 (1991). D. T. Pierce, R. J. Celotta and J. Unguris, J. Appl. Phys. **73**, 6201 (1993).

[19] P. J. H. Bloemen, M. T. Johnson, M. T. H. van der Vorst, R. Coehoorn, J. J. de Vries, R. Jungblut, J. aan de Stegge, A. Reinders and W. J. M. de Jonge, Phys. Rev. Lett. **72**, 764 (1994).

[20] A. Fert, P. Grünberg, A. Barthélémy, F. Petroff and W. Zinn, J. Magn. Magn. Mater. **140-1**, 1 (1995).

[21] Review papers on interlayer couplings for instance are: B. Heinrich and J. A. C. Bland, *Ultrathin Magnetic Films II*, (Spring-Verlag, 1994); P. Grünberg, *Handbook of Magnetic materials*, Vol. 13, edited by K. H. J. Buschow (Elsevier, 1999), p. 1.

[22] F. E. Fullerton, S. D. Bader and J. L. Robertson, Phys. Rev. Lett. **77**, 1382 (1996).

[23] A. Schreyer, J. F. Ankner, T. Zeidler, H. Zabel, M. Schafer. J. A. Wolf, P. Grünberg and C. F. Majkrzak, Phys. Rev. B **52**, 16066 (1995).

[24] K. Mibu, M. Almokhtar, S. Tanaka, A. Nakanishi, T. Kobayashi and T. Shinjo, Phys. Rev. Lett. **84**, 2243 (2000).

[25] D. T. Pierce, J. Unguris, R. J. Celotta and M. D. Stiles, J. Magn. Magn. Mater. **200**, 290 (1999).

[26] T. Shinjo and H. Yamamoto, J. Phys. Soc. Jpn. **59**, 3061 (1990).

[27] K. Shigeto, T. Shinjo and T. Ono, Appl. Phys. Lett. **75**, 2815 (1999).

[28] N. Hosoito, T. Ono, H. Yamamoto, T. Shinjo and Y. Endoh, J. Phys. Soc. Jpn. **64**, 581 (1995).

[29] B. Dieny, V. S. Speriosu, S. S. P. Parkin, B. A. Gurney, D. R. Wilhoit and D. Mauri, Phys. Rev. B **43**, 1297 (1991).

[30] H. Yamamoto, T. Okuyama, H. Dohnomae and T. Shinjo, J. Magn. Magn. Mater. **99**, 243 (1991).

[31] T. Okuyama, H. Yamamoto and T. Shinjo, J. Magn. Magn. Mater. **113**, 79 (1992).

[32] T. Shinjo, *Magnetism and Structure in Systems of Reduced Dimension*, edited by R. F. C. Farrow *et al.*, (Plenum Press, 1995), p. 323; H. Yamamoto, Y. Motomura, T. Anno and T. Shinjo, J. Magn. Magn. Mater. **126**, 437 (1993).

[33] H. Itoh, J. Inoue, and S. Maekawa, J. Magn. Magn. Mater. **126**, 479 (1993).

[34] H. Sato, Y. Aoki, Y. Kobayashi, H. Yamamoto and T. Shinjo, J. Phys. Soc. Jpn. **62**, 431 (1993).

[35] J. Sakurai, M. Horie, S. Araki, H. Yamamoto and T. Shinjo, J. Phys. Soc. Jpn. **60**, 2522 (1991).

[36] J. Shi, R.C. Yu, S.S.P. Parkin and M.B. Salamon, J. Appl. Phys. **73**, 5524 (1993).

[37] K. Nishimura, J. Sakurai, K. Hasegawa, Y. Saito, K. Inomata and T. Shinjo, J. Phys. Soc. Jpn. **63**, 2685 (1994).

[38] E.Yu. Tsymbal, D.G. Pettifor, J. Shi, M.B. Salamon, Phys. Rev. B **59**, 8371 (1999).

[39] W. F. Egelhoff Jr., P. J. Chen, C. J. Powell, M. D. Stiles, R. D. McMichael, C. -L. Lin, J. M. Sivertsen J. H. July, K. Takano, A. E. Berkowitz, T. C. Anthony and J. A. Brug, J. Appl. Phys. **82**, 6142 (1997).

[40] For example, Y. Sugita, Y. Kawawake, M. Satomi and H. Sakakima, Jpn. J. Appl. Phys. **37**, 5984 (1998).

[41] M. Yoshikawa, M. Takagishi, H. Yuasa, K. Koi, H. Iwasaki and M. Sahashi, to be published in Appl. Phys. Lett. **80** (2002).

[42] A. E. Berkowitz, J. R. Mitchell, M. J. Carey, A. P. Young, S. Zhang, F. E. Spada, F. T. Parker, A. Hutten and G. Thomas, Phys. Rev. Lett. **68**, 3745 (1992).

[43] J. Q. Xiao, J. S. Jiang and C. L. Chien. Phys. Rev. Lett. **68**, 3749 (1992).

[44] S. A. Makhlouf, K. Sumiyama, K. Wakoh, K. Suzuki, K. Takanashi and H. Fujimori, J. Magn. Magn. Mater. **126**, 485 (1993).

[45] Ch. Somsen, M. Acet, G. Nepecks and E. F. Wassermann, J. Magn. Magn. Mater. **208**, 191 (2000).

[46] T. Miyazaki and N. Tezuka, J. Magn. Magn. Mater. **139**, L231 (1995). T. Miyazaki, N. Tezuka, S. Kumagai, Y. Ando, H. Kubota, J. Murai, T. Watabe and M. Yokota, J. Phys. D: Appl. Phys. **31**, 630 (1998).

[47] J. S. Moodera, L. R. Kinder, T. M. Wong and R. Meservey, Phys. Rev. lett. **74**, 3273 (1995).

[48] S. S. P. Parkin, this volume.

[49] S. Mitani, H. Fujimori and S. Ohnuma, J. Magn. Magn. Mater. **165**, 141 (1997).

[50] L. Jirman, V. Sechovsky, L. Havela, W. Ye, T. Takabatake, H. Fujii, T. Suzuki, T. Fujita, E. Bruck and F. R. De Boer, J. Magn. Magn. Mater. **104-107**, 19 (1992).

[51] A. Urushibara, Y. Morimoto, T. Arima, A. Asamitsu, G. Kido and Y. Tokura, Phys. Rev. B **51**, 14103 (1995).

[52] *Colossal Magnetoresistive Oxides* (Advances in Condensed Matter Science. Vol. 2), edited by Y. Tokura, (Gordon and Breach Science Publishers, 2000).

[53] S. Maekawa, J. Inoue and H. Itoh, J. Appl. Phys. **79**, 4730 (1996).

[54] J. Mathon, Phys. Rev. B **55**, 960 (1997).

[55] K. Takanashi, S. Mitani, K. Himi and H. Fujimori, Appl. Phys. Lett. **72**, 737 (1998).

[56] M. A. M. Gijs and G. E. W. Bauer, Adv. Phys. **46**, 285 (1997).

[57] W. P. Pratt Jr., S.-F. Lee, J. M. Slaughter, R. Loloee, P. A. Schroeder and J. Bass, Phys. Rev. Lett. **66**, 3060 (1991).

[58] M. A. M. Gijs, S. K. J. Lenczowski and J. B. Giesbers, Phys. Rev. Lett. **70**, 3343 (1993).

[59] M. A. M. Gijs, J. B. Giesbers, S. K. J. Lenczowski and H. H. J. M. Janssen, Appl. Phys. Lett. **63**, 111 (1993).

[60] I. Nakatani, Materia **35**, 854 (1996) (in Japanese).

[61] M.C. Cyrille, S. Kim, M.E. Gomez, J. Santamaria, K.M. Krishnan, and I.K. Schuller, Phys. Rev. B **62**, 3361 (2000).

[62] L. Piraux, J. M. George, J. F. Despres, C. Leroy, E. Ferain, R. Legras, K. Ounadjela and A. Fert, Appl. Phys. Lett. **65**, 2484 (1994); A. Fert and L. Piraux, J. Magn. Magn. Mater. **200**, 338 (1999).

[63] T. Ono and T. Shinjo, J. Phys. Soc. Jpn. **64**, 363 (1995); T. Ono, Y. Sugita, K. Shigeto, K. Mibu, N. Hosoito and T. Shinjo, Phys. Rev. B **55**, 14457 (1997).

[64] M. A. M. Gijs, M. T. Johnson, A. Reinders, P. E. Huisman, R. J. M. van der Veerdonk, S. K. J. Lenczowski and R. M. J. van Gansewinkel, Appl. Phys. Lett. **66**, 1839 (1995).

[65] A. Encinas, F. Nguyen Van Dau, M. Sussiau, A. Schuhl, and P. Galtier, Appl. Phys. Lett. **71**, 3299 (1997).

[66] P. S. Anil Kumar, R. Jansen, O. M. J. van't Erve, R. Vlutters, P. de Haan and J. C. Lodder, J. Magn. Magn. Mater. **214**, L1 (2000).

[67] K. Mizushima, T. Kinno, K. Tanabe and T. Yamauchi, Phys. Rev. B **58**, 4660 (1998).

[68] A text book for MQT phenomena is for instance: E. M. Chudnovsky and J. Tejada, *Macroscopic Quantum Tunneling of the Magnetic Moment* (Cambridge Univ. Press. 1998). References are cited therein.

[69] Examples of recent experimental papers on the resistance of domain walls are; J. F. Gregg, W. Allen, K. Ounadjela, M. Viret, M. Hehn, S. M. Thompson and J. M. D. Coey, Phys. Rev. Lett. **77**, 1580 (1996); K. Hong and N. Giordano, Condens. Matter. **10**, L401 (1998). Y. Otani, S. G. Kim, K. Fukamichi, O. Kitakami and Y. Shimada, IEEE Trans. Mag. **34**, 1096 (1998); K. Mibu, T. Nagahama and T. Shinjo, Phys. Rev. B **58**, 6442 (1998); U. Ruediger, J. Yu, S. Zhang, A. D. Kent and S. S. P. Parkin, Phys. Rev. Lett. **80**, 5639 (1998); Theoretical papers are; P. M. Levy and S. Zhang, Phys. Rev. Lett. **79**, 5110 (1997); G. Tatara and H. Fukuyama, Phys. Rev. Lett. **78**, 3773 (1998).

[70] T. Ono, Y. Sugita and T. Shinjo, J. Phys. Soc. Jpn. **65**, 3021 (1996).

[71] Y. Sugita, T. Ono and T. Shinjo (unpublished).

[72] T. Ono, H. Miyajima, K. Shigeto ad T. Shinjo, Appl. Phys. Lett. **72**, 1116 (1998).

[73] T. Ono, H. Miyajima, K. Shigeto, K. Mibu, N. Hosoito and T. Shinjo, J. Appl. Phys. **85**, 6181 (1999).

[74] K. Shigeto, T. Okuno, T. Shinjo, Y. Suzuki and T. Ono, J. Appl. Phys. **88**, 6636 (2000).

# Chapter 2

## Theory of Giant Magnetoresistance

*Peter M. Levy and Ingrid Mertig*

## 2.1 Introduction

As we have read in the introductory chapter magnetoresistance (MR) is the change in the electrical resistivity $\rho$ or conductivity $\sigma = \rho^{-1}$ of a material when it is subjected to a magnetic field. Ordinarily resistivity increases with field for homogeneous systems; this is known as *positive* MR. There have been examples of *negative* MR, iron whiskers, where the resistivity decreases with field. One reason for this comes from the field erasing domain walls; they are a source of additional resistance, beyond that in the bulk of the domains themselves. However in 1988, a new class of magnetic structures were grown, metallic multilayers; these consisted of magnetic layers separated by non-magnetic ones, e.g., Fe/Cr and Co/Cu. If in zero field the adjacent magnetic layers are antiparallel (AP) or randomly oriented they displayed large decreases in their resistance as an external magnetic field oriented the moments of the magnetic layers in parallel (P). This large decrease is called *giant magnetoresistance* (GMR); at low temperatures it is typically of the order of 50% [1], and has been found to be as large 220% [2]. GMR is usually quoted as the ratio of the difference in the resistivities $\rho$ of a magnetic multilayer in zero magnetic field and the field needed to saturate the magnetization, to the resistivity at saturation. If one posits that the magnetic configuration in zero field is antiferromagnetic this ratio is

$$R = \frac{\rho_{AP} - \rho_P}{\rho_P} = \frac{\sigma_P}{\sigma_{AP}} - 1;  \tag{2.1}$$

this is the definition adopted in the introductory chapter. In the normal GMR effect $\rho_{AP} > \rho_P$ and $R$ is unbounded. Another definition is

$$R' = \frac{\rho_{AP} - \rho_P}{\rho_{AP}} = 1 - \frac{\sigma_{AP}}{\sigma_P} \qquad (2.2)$$

which is bounded $0 < R' < 1$. The zero field state of a magnetic multilayer is not always well defined, i.e., it may be a random configuration for which the net magnetization is zero, therefore most people prefer to use $R$. Parenthetically, it gives a larger value of the ratio, as long as $\rho_{AP} > \rho_P$. The inverse GMR effect has also been found in which $\rho_P > \rho_{AP}$, and is described in a recent summary on this subject by Vouille *et al.* [3]. In this chapter, we describe the formalism for understanding transport in magnetic multilayers, review the origins of GMR, discuss the results to date, and indicate the relation of the MR found in magnetic tunnel junctions, Sec. 2.6, to that in metallic multilayers.

To understand the magnetoresistive properties of magnetic multilayers we first review the different formalisms for calculating electrical transport in solids; then we adapt them to magnetic systems. The two current model is a simple explanation of the conduction process peculiar to magnetic structures. For inhomogeneous structures we have at least three levels of problems to address. First the electronic band structure and the bare conductivity tensor it yields; second the distribution of impurity scatterers, and how one does the appropriate impurity averaging to calculate the decay of electron propagators; and finally how the orientation of the externally applied electric field relative to the layering in multilayered structures creates non-equilibrium (current driven) charge accumulation. For magnetic materials we include the spin-dependence of the band structure and scattering, and the current driven magnetization. As it is quite hard to determine current driven charge and spin accumulations for arbitrary directions of the electric field relative to the layering, we focus on two principal transport geometries for magnetic multilayers: current in the plane of the layers (CIP), and current perpendicular to the plane of the layers (CPP); see Fig. 1.17 of the introductory chapter.

## 2.1.1   Simple concepts of transport

Electrical resistance will be traced to the loss of momentum information of electrons; the Drude formula

$$\varrho = \frac{m}{ne^2\tau} \qquad (2.3)$$

is perhaps the simplest formula that relates resistivity to the mass $m$ and charge $e$ of an electron, the number of electrons per unit volume $n$ and the time between collisions $\tau$ [4]. In order to better bring out the essential material parameters that enter resistivity, it can be written as the inverse of the conductivity [5]

$$\sigma = e^2 \sum_k \delta(E_k - E_F) v_k \Lambda_k. \qquad (2.4)$$

We see the conductivity is a sum over the Fermi surface of the velocity (actually the component along the direction of the electrical field) $v_k$ and mean free path $\Lambda_k = \tau_k v_k$ of the conduction electrons, where $\tau_k$ is the relaxation time. As $\sum_k \delta(E_k - E_F)$ defines the density of states $N(E_F)$ at the Fermi energy, the conductivity or resistivity at low temperatures ($T \approx 0\,\mathrm{K}$) is controlled by three factors: the density of states, velocity, and relaxation time of the electrons on the Fermi surface. The first two are completely determined by the electronic structure of the solid; the last one is determined by the defects or impurities present in the structure. This expression can be reduced to the Drude formula for the conductivity by assuming an isotropic relaxation time $\tau$ and writing

$$\sum_k \delta(E_k - E_F) v_k^2 = \frac{2}{3m} \sum_k \delta(E_k - E_F) E_k = \frac{2}{3m} N(E_F) E_F = \frac{n}{m}, \qquad (2.5)$$

where we approximated on average $v_k^2$ as $\frac{1}{3} v_k^2$ as it represents the component of the velocity in one spatial direction for isotropic systems.

While the above simplifications are reasonable for homogeneous systems they are more often than not inadequate for layered structures. Most prominently, the relaxation time is not isotropic; specifically the vector mean free path that enters the proper expression for conductivity of a multilayer structure depends on the direction of the $k$ vector and it is not simply related to the relaxation time [5]. Due to screening in metals electronic structure is determined by local environment; as screening lengths are typically of the order of 1-3 Å in good metals the electronic structure about an atomic site is controlled by the nearest 3-4 neighbor shells of atoms. The extent to which local variations in a multilayer affect transport is determined by the size of the region over which electrons retain a memory of their momentum direction, i.e., the mean free path; see Fig. 2.5 later on. This length scale depends on strength of impurity scattering and their concentration.

Without impurities and at zero temperature, electrons maintain knowledge of their momentum direction; this does not imply that they do not undergo specular scatterings which reorient the momentum, but they do so in a predictable and reproducible fashion. This is known as the *ballistic* regime, in which one determines the electronic structure and transport for the entire solid. Impurities introduce additional scattering; if their number is so small that one can solve the quantum mechanical problem of the electron states in their presence there is no loss of memory of momentum and in some sense no resistance. In this case, one talks of the conductance of the structure which is related to the transmission probabilities for electrons traveling between two leads (reservoirs) by the Landauer-Büttiker formalism [6]. It is indeed rare that electrons can travel across a macroscopic sample and encounter only a few impurities; it is definitely not the case for the magnetic multilayers in which GMR has been observed. Nonetheless, in such situations electrons travel in eigenstates specific to the region as long as they do not suffer inelastic collisions, i.e., remain coherent. Therefore, one solves for electron states in a region determined by the coherence length of the electrons of the structure for a specific realization of a fixed density of impurities, and calculates the conductance. At very low temperatures the coherence length is so large that it is impractical to determine the eigenstates for the number of impurities encompassed, and one must truncate the region to evaluate. Under the influence of an electric field electrons travel further down the structure and enter regions with different realizations of impurity distributions, albeit with the same density of impurities. To account for the different scattering situations electrons experience in traveling across a macroscopic conductor one must repeat the calculations of the eigenstates and conductances for all different realizations of impurity distributions, and average the conductances to obtain an average conductivity (conductance per unit length) which is the experimentally measured quantity.

### 2.1.2   Length scales for inhomogeneous systems

The calculation of the conductivity can be considerably simplified for macroscopic conductors by introducing the concept of impurity averaged propagators; in this picture the effect of impurities on electron states is accounted for by a fictitious potential which has the same symmetry as the underlying potential of the periodic structure [7]. While electrons are not scattered from one $k$ state to another by this potential they do decay, so that one immediately has introduced $\Lambda_k$ or $\tau_k$, mean free paths or relaxation times, in which electrons travel in a well-defined momentum states for a finite distance or time. In this picture, one introduces loss of momentum direction memory from the start and resistivity is readily related to the imaginary part of the self energy coming from the fictitious potential representing the averaged impurity scattering. Properly defined the conductivity of a macroscopic sample calculated by averaging the conductance of regions with different realizations of the impurity distribution, and that found by averaging the impurity scattering potential are the same. When the concentration of impurities necessitate using impurity averaged propagators, one is in the regime of *diffusive* transport.

The transport in the GMR devices studied to date is diffusive, so that it is easiest to discuss their transport properties in terms of impurity averaged propagators. In addition, they are inhomogeneous structures, e.g., the local density of states (LDOS) and the density and nature of the impurities are different in the middle of a layer from that at the interface. Therefore, the above procedure for calculating the resistivity has to be modified, interalia, we must introduce new length scales for evaluating resistivity, and as we will see they will depend on the direction of the current relative to the layering in the structure. We will assume that within an atomic plane, monolayer (ML), the concentration of impurities is constant, and that it varies along the growth direction of the multilayer, i.e., normal to the plane of the layers. In principle each atomic plane has different concentrations and types of impurities.

## 2.2 Formalisms

The problem under consideration is a system of interacting electrons moving in the electrostatic potential caused by the nuclei. By means of density functional theory [8] this general problem can be transformed to an effective one-electron Hamiltonian

$$H_o = \frac{p^2}{2m} + V_{pot,M}^{\sigma}(\mathbf{r}) \tag{2.6}$$

consisting of the kinetic energy and the effective one-electron potential. For magnetic multilayers the potential is inhomogeneous due to the layered structure. Here $V_{pot,M}^{\sigma}(\mathbf{r})$ represents the spin dependent potentials of the electrons in the different regions of a multilayered structure and M denotes the magnetic configuration, e.g., ferromagnetic (P), antiferromagnetic (AP), or random. The corresponding one-electron Schrödinger equation

$$H_0\varphi_0(k,\mathbf{r},E) = E_k\varphi_0(k,\mathbf{r},E) \tag{2.7}$$

will give us the one-electron energies $E_k$ and eigenfunctions, Bloch functions $\varphi_0(k,\mathbf{r},E)$ characterizing the system. Here $k$ is a shorthand notation for $(\nu,\mathbf{k},\sigma)$, where $\nu$ is the band index, $\mathbf{k}$ is momentum, and $\sigma$ spin.

As discussed in the introduction, a translationally invariant system would not cause any resistivity. For this reason, we have to consider a perturbed system

$$H = H_0 + V_{scatt} \tag{2.8}$$

incorporating some scattering potential

$$V_{scatt} = \sum_a (v_a + j_a\hat{\mathbf{M}}_a \cdot \hat{\sigma})\delta(\mathbf{r} - \mathbf{r}_a) \tag{2.9}$$

where $\hat{\sigma}$ stands for the Pauli spin vector operator and $\hat{\mathbf{M}}_a$ represents the orientation of the magnetization of the corresponding region. For simplicity we have taken the range of this scattering potential to be zero; this simplifies the discussion of the local scattering without altering the effect we are modeling. For the non-magnetic regions $j = 0$ (see Eq. (2.9)), while at interfaces and in magnetic regions (granules or layers) $j \neq 0$.

## 2.2.1   Boltzmann

For the description of transport properties of a metallic system from a microscopic level one can take advantage of the quasi-classical approach of transport
based on the Boltzmann equation [4, 9]. This description accounts for the microscopic origin of the transport coefficients and is valid as long as the mean
free path is large as compared to the lattice constant and small relative to the
macroscopic dimension of the system.

The Boltzmann theory assumes the existence of a distribution function
$f_k(\mathbf{r})$ which measures the number of carriers in the state $k$ in the neighborhood
of $\mathbf{r}$. This distribution function changes through diffusion of charge carriers,
under the influence of external fields and due to scattering, giving rise to a
rate of change

$$\frac{\partial f_k(\mathbf{r})}{\partial t}$$

which should vanish in the steady state. The resulting condition is the Boltzmann equation

$$\left(\frac{\partial f_k(\mathbf{r},t)}{\partial t}\right) + \left(\frac{\partial f_k(\mathbf{r})}{\partial t}\right)_{diffusion} + \left(\frac{\partial f_k(\mathbf{r})}{\partial t}\right)_{field} = -\left(\frac{\partial f_k(\mathbf{r})}{\partial t}\right)_{scatt} \qquad (2.10)$$

where the terms at the left hand side account for changes of the distribution
function due to: an explicit time dependence, diffusion and influence of external fields. All these changes have to be in equilibrium with the changes of the
distribution function through scattering.

We restrict the following considerations to a homogeneous external electric
field $\mathbf{E}$ in which case only the field term exists and Eq. (2.10) reduce to

$$e\left(\frac{\partial f^o_k}{\partial E_k}\right)\mathbf{v}_k\mathbf{E} = \left(\frac{\partial f_k}{\partial t}\right)_{scatt} \qquad (2.11)$$

where $\mathbf{v}_k$ is the velocity of the electrons which can be obtained from the single-
particle energies of the considered system

$$\mathbf{v}_k = \frac{1}{\hbar}\frac{\partial E_k}{\partial \mathbf{k}}, \qquad (2.12)$$

$f^o_k$ is the Fermi-Dirac distribution function

$$f^o_k = \frac{1}{e^{(E_k-E_F)/k_B T} + 1} \qquad (2.13)$$

at $T = 0\,\mathrm{K}$, and $k_B$ is Boltzmann's constant. Here we use the convention that
the $e$ is the absolute value of the charge of the electron. The sign is already
incorporated in the equation. The local change of charge carriers due to elastic

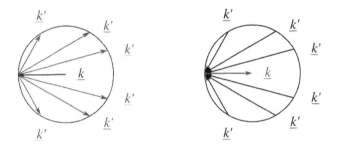

Scattering-out term          Scattering-in term

Figure 2.1: Scattering-out term (left) and scattering-in term (right) as visualized on the Fermi sphere. Note that in the first an electron comes in at one **k** vector and is scattered to all states on the Fermi surface, while in the second the reverse happens.

scattering of independent particles can be related to the microscopic scattering probability $P_{kk'}$ by

$$\left(\frac{\partial f_k}{\partial t}\right)_{scatt} = \sum_{k'} [f_{k'}(1-f_k)P_{k'k} - (1-f_{k'})f_k P_{kk'}]. \tag{2.14}$$

The first term characterizes the scattering of electrons from occupied states $k'$ into an empty state $k$ (scattering-in term or vertex correction term) while the second term characterizes the reverse process, the scattering of an electron from an occupied state $k$ into empty states $k'$ (scattering-out term). The meaning of the two contributions is illustrated in Fig. 2.1. We separate the distribution function $f_k$ into two parts

$$f_k = f_k^o + g_k, \tag{2.15}$$

where the deviations $g_k$ from the equilibrium distribution function are assumed to be modest under a small external electric field **E**. According to the principle of microscopic reversibility

$$P_{kk'} = P_{k'k} \tag{2.16}$$

and in view of Eq. (2.15) the ansatz for the local change of charge carriers (Eq. (2.14)) simplifies to

$$\left(\frac{\partial f_k}{\partial t}\right)_{scatt} = \sum_{k'} P_{kk'}(g_{k'} - g_k). \tag{2.17}$$

By assuming a linear response for the deviations of the electron distribution function we know that $g_k \sim |\mathbf{E}|$ and make the ansatz

$$g_k = -e \left( \frac{\partial f_k^o}{\partial E_k} \right) \mathbf{\Lambda}_k \cdot \mathbf{E}, \tag{2.18}$$

where $\mathbf{\Lambda}_k$ is the *vector* mean free path. The magnitude of the vector mean free path is the usual mean free path $\Lambda$ (see also Eq. (2.4)) known from text books and measures the path of the electron between two successive scattering events. With Eqs. (2.17) and (2.18) the Boltzmann equation (Eq. (2.11)) can be transformed into an equation for the vector mean free path [5]

$$\mathbf{\Lambda}_k = \tau_k \left[ \mathbf{v}_k + \sum_{k'} P_{kk'} \mathbf{\Lambda}_{k'} \right] \tag{2.19}$$

Here $\tau_k$ is the relaxation time defined by

$$\tau_k^{-1} = \sum_{k'} P_{kk'}; \tag{2.20}$$

it characterizes the scattering process most meaningfully for it represents the time that an electron stays in the state $k$ until the next scattering event. It has to be mentioned that as it is defined here $\tau$ is a life time and has to be distinguished from the transport relaxation time $\tau_{tr}$ which includes a factor $(1 - \cos\theta_{kk'})$ due to the scattering in term [9].

In non-magnetic systems Eq. (2.19) is an integral equation from which the vector mean free path $\mathbf{\Lambda}_k$ $(k = (\nu, \mathbf{k}))$ can be determined. In ferromagnetic systems the scattering probability $P_{kk'}$ is described by a $2 \times 2$ matrix in spin space, i.e., it contains four components depending upon the spin quantum numbers $\sigma, \sigma'$ before and after scattering (Fig. 2.2)

$$P_{kk'}^{\sigma\sigma'} = \begin{pmatrix} P_{kk'}^{\uparrow\uparrow} & P_{kk'}^{\uparrow\downarrow} \\ P_{kk'}^{\downarrow\uparrow} & P_{kk'}^{\downarrow\downarrow} \end{pmatrix}. \tag{2.21}$$

As a consequence, the Boltzmann equation becomes a system of coupled integral equations which decouples if spin-flip scattering processes are ignored. Parenthetically, integrations are performed over the anisotropic Fermi surface of the system under consideration.

The current density is written

$$\mathbf{j} = e \sum_k f_k \mathbf{v}_k \tag{2.22}$$

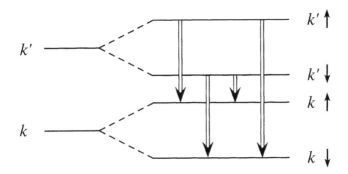

Figure 2.2: Spin-dependent scattering in ferromagnetic systems. There are transitions that conserve spin $k'^\uparrow \to k^\uparrow$, as well as those which involve spin-flips $k'^\uparrow \to k^\downarrow$.

which is the product of charge density along a velocity direction and the velocity of charge carriers. The crystal volume is set to unity. By combining Eqs. (2.15), (2.18) and (2.22) and comparing the result to Ohm's law we find

$$\mathbf{j} = \overleftrightarrow{\sigma} \mathbf{E} \qquad (2.23)$$

where the conductivity tensor $\overleftrightarrow{\sigma}$, or the inverse resistivity tensor $\overleftrightarrow{\rho}^{-1}$, is a dyadic ($3 \times 3$ tensor) in real space

$$\overleftrightarrow{\rho}^{-1} = \overleftrightarrow{\sigma} = e^2 \sum_k \delta(E_k - E_F)\, \mathbf{v}_k \mathbf{\Lambda}_k. \qquad (2.24)$$

The components of the conductivity tensor are equal in cubic systems which is assumed in Eqs. (2.3), (2.4), and (2.5). In systems with lower symmetry the components of the conductivity tensor differ along the symmetry axes. This is known as resistivity anisotropy in non-cubic systems and also shows up in layered structures since they have only tetragonal symmetry. As a result the conductivity can be measured with CIP or CPP of the layers.

In ferromagnetic systems where the spin degeneracy is lifted, the conductivity is usually split into spin-dependent contributions

$$\overleftrightarrow{\sigma}^\sigma = e^2 \sum_k \delta(E_k^\sigma - E_F)\mathbf{v}_k^\sigma \mathbf{\Lambda}_k^\sigma \qquad (2.25)$$

where $\overleftrightarrow{\sigma}^\uparrow$ and $\overleftrightarrow{\sigma}^\downarrow$ stand for the majority and the minority electrons, respectively. Consequently, the factor 2 in Eq. (2.24) disappears and the total conductivity is given by the sum of both contributions due to the two current model [10]

$$\sigma = \sigma^\uparrow + \sigma^\downarrow; \qquad (2.26)$$

here we dropped the tensor notation, and thereby assumed we are taking one component of the conductivity tensor. Usually, this is the one parallel to the plane of the layers in multilayered structures, but we also take the component perpendicular to the layers. The spin anisotropy ratio

$$\alpha = \sigma^\uparrow/\sigma^\downarrow = \rho^\downarrow/\rho^\uparrow \qquad (2.27)$$

is usually introduced [11] to account for the individual contributions of the majority and minority electrons to the transport properties.

Since the solution of Eq. (2.19) is a difficult problem several approximations are made. If the scattering-in term of Eq. (2.19) is neglected and a constant relaxation time $\tau$ assumed the conductivity becomes

$$\overleftrightarrow{\sigma}^\sigma = e^2 \tau \sum_k \delta(E_k^\sigma - E_F)\, \mathbf{v}_k^\sigma \mathbf{v}_k^\sigma. \qquad (2.28)$$

This expression is equivalent to Eq. (2.3) for cubic symmetry. The isotropic relaxation time assumes that all electrons undergo the same scattering; it can be considered as an average over the Fermi surface of all state dependent relaxation times

$$\tau = \frac{\sum_k \delta(E_k - E_F)\, \tau_k}{\sum_k \delta(E_k - E_F)}. \qquad (2.29)$$

This simple model does not adequately describe scattering in magnetic multi-layers. As a minimum the spin-dependence of scattering has to be taken into account in magnetic systems, that is, we consider different relaxation times for the majority $\tau^\uparrow$ and minority electrons $\tau^\downarrow$ to account for the individual contributions of the majority and minority electrons to the conductivity

$$\overleftrightarrow{\sigma}^\sigma = e^2 \tau^\sigma \sum_k \delta(E_k^\sigma - E_F)\, \mathbf{v}_k^\sigma \mathbf{v}_k^\sigma; \qquad (2.30)$$

this is accompanied by the spin anisotropy ratio

$$\beta = \tau^\uparrow/\tau^\downarrow. \qquad (2.31)$$

More sophisticated ab-initio calculations [12] (see Sec. 2.5.3.5) have shown that the relaxation times in magnetic multilayers are highly anisotropic. This anisotropy has to be taken into account

$$\overleftrightarrow{\sigma}^\sigma = e^2 \sum_k \delta(E_k^\sigma - E_F)\, \tau_k^\sigma\, \mathbf{v}_k^\sigma \mathbf{v}_k^\sigma \qquad (2.32)$$

to obtain quantitatively reliable results.

## 2.2.2 Landauer

The Landauer-Büttiker scattering theory [6, 7] of transport is an established method for describing transport in mesoscopic systems where the system size plays an important role in determining the conductances. In applying this formalism, assumptions are made about the measuring configuration, i.e., about the leads and contacts which inject and drain the electric current. For a general band structure the Landauer conductance formula can be written in terms of incoming and outgoing Bloch states that are labeled by $k_\parallel$

$$G(\mathbf{n}) = \frac{e^2}{h} \sum_{k_\parallel, k_\parallel'} \left| T_{k_\parallel, k_\parallel'} \right|^2 \tag{2.33}$$

which is a short hand notation for $(\nu, \mathbf{k}_\parallel, \sigma)$ where $\mathbf{k}_\parallel$ is the component of the Bloch vector perpendicular to the current direction $\mathbf{n}$. The calculation of the transmission probabilities $|T_{k_\parallel, k_\parallel'}|^2$ from the incident mode $k_\parallel$ to the transmitted mode $k_\parallel'$ is in general difficult. In the ballistic regime, however, this calculation is trivial because the modes are not scattered at all and the transmission probability matrix is simply $\delta_{k_\parallel, k_\parallel'}$. Then the conductance becomes

$$G(\mathbf{n}) = \frac{e^2}{h} \sum_{k_\parallel, k_\parallel'} \delta_{k_\parallel, k_\parallel'} \tag{2.34}$$

which is proportional to the number of conducting channels for transport in direction of the current. By transforming Eq. (2.34) for a 3D electron gas into a Fermi surface integral we obtain

$$G(\mathbf{n}) = \frac{e^2}{h} \frac{A}{4\pi^2} \frac{1}{2} \sum_{k} \delta(E_k - E_F) \left| \mathbf{n} \cdot \mathbf{v}_k \right| \tag{2.35}$$

where $A/4\pi^2$ is the density of transverse modes (like the factor $V/8\pi^3$ for volume integrations in reciprocal space) and $\mathbf{n} \cdot \mathbf{v}_k$ is the component of the Fermi velocity in direction of the current. The factor $1/2$ in Eq. (2.35) appears because only electrons moving in direction of the current contribute. The expression (Eq. (2.35)) is known as the ballistic or Sharvin conductance [13]. The corresponding current along the transport direction $\mathbf{n}$ is given by $I = VG(\mathbf{n})$. The conductance given in Eq. (2.35) describes transport through a ballistic point contact which consists of two electrodes connected via a narrow region. The device is referred to as a ballistic point contact when the diameter of the narrow region is much smaller than the mean free path and much larger than the electron wavelength. The conductance of such a point contact is determined by the ballistic motion of the electrons through the region. Even though the electrons passing through the constriction are not scattered, the conductance of a point contact is finite due to its' finite cross-section A. Although this experimental situation is not realized in most of experiments measuring GMR, the formalism was applied and discussed by several authors [14, 15] to elucidate the microscopic origin of GMR. A comparison of the spin-projected contributions of the conductance

$$G^\sigma(\mathbf{n}) = \frac{e^2}{h} \frac{A}{4\pi^2} \frac{1}{2} \sum_k \delta(E_k^\sigma - E_F) \, |\mathbf{n}\cdot\mathbf{v}_k^\sigma|, \tag{2.36}$$

with the spin-projected conductivity in the relaxation time approximation (Eq. (2.28)) manifests the similarities. Both the conductance and the conductivity are determined by a Fermi surface integral of either the absolute value or the square of the Fermi velocity along the current direction. The consequences of this analogy will be discussed in Sec. 2.5.1.

### 2.2.3   Kubo

In the 1950s, Kubo developed a method of evaluating the response of a quantum mechanical system to an external potential; in particular the current in response to an electric field [16]. To first order, known as linear response, the two are related by a conductivity which is given in terms of the equilibrium properties of the system, i.e., in zero field.

For non-magnetic (normal) metals the current at a point $\mathbf{r}$ is related to the static electric field at a point $\mathbf{r}'$ through the two-point conductivity

$$\mathbf{j}(\mathbf{r}) = \int d^3 r' \overleftrightarrow{\sigma}(\mathbf{r}, \mathbf{r}') \cdot \mathbf{E}(\mathbf{r}') \tag{2.37}$$

where $\mathbf{E}(\mathbf{r}')$ is the electric field in the solid and $\overleftrightarrow{\sigma}(\mathbf{r}, \mathbf{r}')$ is the two-point conductivity given by Kubo's linear-response formalism which in the zero-frequency limit is [17, 18]

$$\overleftrightarrow{\sigma}(\mathbf{r}, \mathbf{r}') = \lim_{\omega\to 0} \frac{\overleftrightarrow{\Pi}(\mathbf{r}, \mathbf{r}'; \omega)}{\omega} \tag{2.38}$$

where $\overleftrightarrow{\Pi}(\mathbf{r}, \mathbf{r}'; \omega)$ is the frequency-dependent current-current correlation function,

$$\overleftrightarrow{\Pi}(\mathbf{r}, \mathbf{r}'; \omega) = \int_0^\infty d\tau e^{i\omega\tau} \left\langle \left[ \mathbf{j}_{op}(\mathbf{r}, t+\tau), \mathbf{j}_{op}(\mathbf{r}', t) \right] \right\rangle \tag{2.39}$$

and $\mathbf{j}_{op}(\mathbf{r}, t)$ is the quantum-mechanical current operator

$$\mathbf{j}_{op}(\mathbf{r}, t) = e^{iH_o t} \mathbf{j}_{op}(\mathbf{r}) e^{-iH_o t} \tag{2.40}$$

$$\mathbf{j}_{op}(\mathbf{r}) = \frac{e\hbar}{mi} \Psi^\dagger(\mathbf{r}) \overleftrightarrow{\nabla}_r \Psi(\mathbf{r}) \tag{2.41}$$

where

$$\overleftrightarrow{\nabla}_r \equiv \frac{1}{2}(\overrightarrow{\nabla}_r - \overleftarrow{\nabla}_r) \tag{2.42}$$

is the antisymmetric gradient operator and $\Psi(\mathbf{r})$ is the real space one electron field operator. The double arrows on this *vector* operator should not be confused with those on the conductivity *tensor*. The correlation function is independent of $t$ due to time-translation invariance. The angular brackets denote the expectation value of the commutator taken over all states of the system (we limit ourselves to zero temperature). Therefore, to calculate the conductivity it is necessary to start from the Hamiltonian that describes the conduction electrons (Eq. (2.8)). Also we are primarily interested in transport at $T = 0\,\mathrm{K}$, so that processes that occur at finite temperature are omitted.

Electrical transport in metals leads to a redistribution of space charge if the rate at which the electrons are scattered varies from one region to another. For this reason, the local or internal field is not the same as the field applied externally; i.e., while one might apply a uniform electric field, the internal field is not uniform in inhomogeneous media. Transport in magnetic metals with spin dependent potentials and scattering leads to a spatial redistribution of spin as well as charge; this is referred to either as spin accumulation or non-equilibrium (current-driven) magnetization. To account for this spin-polarized conduction [see Eqs. (2.6) and (2.9)], it is necessary to introduce the spin variables referring to the conduction electrons in the constitutive relation that relates fields to currents. This program can be implemented by defining the generalized spin-dependent current densities which are expectation values of the spinor current operators [19]

$$\mathbf{j}_{op}^{\alpha\beta}(\mathbf{r}) = \frac{e\hbar}{mi}\Psi_\alpha^\dagger(\mathbf{r})\overleftrightarrow{\nabla}_r\Psi_\beta(\mathbf{r}) \tag{2.43}$$

where $\Psi_\beta(\mathbf{r})$ is a component a spinor wavefunction and the Greek indices label the two spin states of the electron: $\uparrow$ and $\downarrow$. Here we limit our discussion to collinear spin structures and neglect spin-flip scattering processes; in this case we have conduction by two *independent* spin channels, and we treat each separately, i.e., $\mathbf{j}^{\alpha\beta}(\mathbf{r}) = \mathbf{j}^\sigma(\mathbf{r})\delta_{\alpha\sigma}\delta_{\beta\sigma}$. We will for the most part suppress the spin index in the following discussion.

By introducing the spinor notation above, we find that the spinor current at point 1 is related to the external electric field at point 3 by [19]

$$\mathbf{j} = \int d3\,\overleftrightarrow{\sigma}(13) \cdot \mathbf{E}_{ext}(3)$$

$$= -\frac{4}{\pi}\frac{e^2}{\hbar}\left(\frac{\hbar^2}{2m}\right)^2 \int d2d3d4 \left\{ G(12)\overleftrightarrow{\nabla}\overleftrightarrow{\Gamma}(234)G(41) \right\} \cdot \mathbf{E}_{ext}(3), \tag{2.44}$$

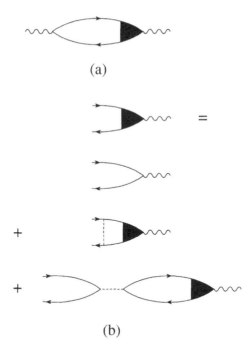

Figure 2.3: (a) Feynman diagram for the conductivity Eq. (2.44) consisting of electron and hole propagators coupled at the right end to an electric field through a *dressed* vertex. (b) This shaded portion represents the vertex correction Eq. (2.45); this term consists of three parts: the bare vertex, the corrections due to interactions between electrons, and those coming from the impurity averaging, that couple the electron and hole propagators; and polarization corrections which modify the field seen by the electrons; taken from [19].

where the Feynman diagram associated with this expression is given in Fig. 2.3 (a). Here $G(12)$ is a one-electron propagator. In Eq. (2.44), as well as in all subsequent equations, spin indices on the propagators and vertex functions are implied. The vector vertex function $\overleftrightarrow{\Gamma}(234)$ is given by the integral equation [see Fig. (2.3(b))],

$$\overleftrightarrow{\Gamma}(234) = \overleftrightarrow{\nabla}_3 \delta(2,3)\delta(4,3) + V(2,4) \int d2' d4' G(2,2') \overleftrightarrow{\Gamma}(2'34') G(4',4)$$

$$+ \delta(2,4) \int d2' d4' d5 V(2,5) G(5,2') \overleftrightarrow{\Gamma}(2'34') G(4',5), \qquad (2.45)$$

where the $V$'s represent interactions not accounted for in the one-particle propagator $G$, matrix multiplication is implied, and $\delta(1,2)$ stands for the Dirac delta function $\delta(\mathbf{r}_1 - \mathbf{r}_2)$. While the first term represents the interaction of an electron with the external field, the other terms represent two electron (Coulomb) interactions, and the effects of the other electrons that have rearranged themselves in response to the field.

In addition we have to invoke loss of momentum information to define a resistance at zero temperature [20]; here we do this by using impurity averaged one electron propagators $G(1,2)$, i.e., propagators where the actual scattering potential, Eq. (2.9), is replaced by an impurity averaged self-energy. As the conductivity, Eq. (2.44) is given as an impurity averaged two electron propagator, the second term in the vector vertex function, Eq. (2.45) includes vertex corrections to the electron-hole propagators (two-particle Green's functions), which represent correlations due to scattering at different sites that are lost when one uses one particle Green's functions with impurity-averaged self-energies.

By confining corrections in the second term of the vertex function, Eq. (2.45), to those that can be represented by a local field, i.e., where 2 and 4 are one and the same position in space, e.g., to lowest order in the density of scatterers, vertex corrections due to impurity averaging are local, and by using the vertex function, we can define the effective *local* field acting on an electron in response to an external field as follows. First, we define a scalar vertex function [17]

$$\overleftrightarrow{\Gamma}(234) = \overleftrightarrow{\nabla}_2 \Gamma(234) \tag{2.46}$$

where the gradient operates on the Green's functions when it is placed in Eq. (2.45). Then, we write

$$\int d3\Gamma(234)\mathbf{E}_{ext}(3) = \mathbf{E}^\sigma(2)\delta(2,4), \tag{2.47}$$

with the understanding that the effective field $\mathbf{E}(2)$ depends on the spin because the vertex function $\Gamma(234)$ is spin dependent. Upon placing this approximation in Eq. (2.44) we find

$$\mathbf{j}^\sigma(\mathbf{r}) = \int d^3 r' \overleftrightarrow{\sigma}^\sigma(\mathbf{r}, \mathbf{r}') \cdot \mathbf{E}^\sigma(\mathbf{r}'), \tag{2.48}$$

where

$$\overleftrightarrow{\sigma}^\sigma(\mathbf{r}, \mathbf{r}') = -\frac{4}{\pi} \frac{e^2}{\hbar} \left(\frac{\hbar^2}{2m}\right)^2 \left\{ G^\sigma(\mathbf{r}, \mathbf{r}') \overleftrightarrow{\nabla}_r \overleftrightarrow{\nabla}_{r'} G^\sigma(\mathbf{r}', \mathbf{r}) \right\}. \tag{2.49}$$

The propagators entering the conductivity are to be interpreted as Matsubara Green's functions [17]; they lead to combinations of retarded and advanced Green's functions, so that the two-point spinor conductivity is explicitly written as

$$\overleftrightarrow{\sigma}^\sigma(\mathbf{r}, \mathbf{r}') = \frac{4}{\pi} \frac{e^2}{\hbar} \left( \frac{\hbar^2}{2m} \right)^2 \left\{ A^\sigma(\mathbf{r}, \mathbf{r}') \overleftrightarrow{\nabla}_\mathbf{r} \overleftrightarrow{\nabla}_{\mathbf{r}'} A^\sigma(\mathbf{r}', \mathbf{r}) \right\} \tag{2.50}$$

where

$$A^\sigma(\mathbf{r}, \mathbf{r}') = \frac{i}{2} [G^\sigma_{ret}(\mathbf{r}, \mathbf{r}') - G^\sigma_{adv}(\mathbf{r}, \mathbf{r}')] \tag{2.51}$$

is the density of states function for one direction of the spin. This conductivity contains only the contributions from the bubble diagram in the Kubo formula; this is the diagram in Fig. 2.3(a) without vertex corrections, i.e., Fig. 2.3(b) with just a bare vertex. From Eqs. (2.47) and (2.48) it is clear that the charge and spin accumulations attendant to conduction in inhomogeneous magnetic structures are accounted for in the effective internal field, $\mathbf{E}^\sigma(\mathbf{r}')$, which has absorbed the vertex corrections.

Finding the global or measurable conductivity, i.e., the net current density across the boundaries of a sample for an applied potential, for inhomogeneous structures from the local conductivity is by no means straightforward. The main difficulty arises from the lack of symmetry in general, which leads to complicated distributions (magnitudes and directions) for both current densities and internal fields. If one could evaluate the internal field $\mathbf{E}^\sigma(\mathbf{r}')$ from the vertex function, Eq. (2.47), one could find the current density from Eq. (2.48); this is equivalent to solving the conductivity in Eq. (2.44) when the field $E$ is simply related to the applied potential. As we will see except for particularly simple geometries, e.g., CIP, this is not possible. In these cases, we resort to approximations for the vertex function that conserve current, i.e., we assure a steady-state current density [19]. This is the trace over spin space of the spinor current density

$$\mathbf{j}(\mathbf{r}) = Tr[\mathbf{j}(\mathbf{r})] = \sum_\sigma \mathbf{j}^\sigma(\mathbf{r}) \tag{2.52}$$

and states that the total current is the sum of the currents in the two spin channels of the electron; this is what is commonly referred to as the two current model [10, 11]. In the steady state, the spinor current densities satisfy the equation [19]

$$\nabla \cdot \mathbf{j}^\sigma(\mathbf{r}) = w^\sigma(\mathbf{r}). \tag{2.53}$$

The right hand side looks-like charge buildup or loss, however it is due to spin-flip processes; therefore we can define spin-flip processes as those that give rise to finite $w^\sigma(\mathbf{r})$. Conversely when $w^\sigma(\mathbf{r}) = 0$ we say there are no spin-flip processes. It is important to distinguish situations with $w^\sigma(\mathbf{r}) \neq 0$ as they reduce the GMR effect due to spin mixing [11], from $w^\sigma(\mathbf{r}) = 0$, where the steady-state current density is conserved in each spin component, i.e., the two spin channels are *independent*.

When the full conductivity, Eqs. (2.44) and (2.45), is used the equation of continuity is trivially satisfied independently of the electric field distribution $\mathbf{E}(\mathbf{r}')$ [19]. However, the bubble conductivity (2.50) and its equivalent, when taking the spin trace, are not divergenceless. Therefore, imposing the constraint of Eq. (2.53) on the current density given by Eq. (2.48) provides a set of equations that allows one in some cases to determine the effective internal fields namely $\mathbf{E}^\sigma(\mathbf{r}')$,

$$\int d^3r' \nabla_\mathbf{r} \cdot \overleftrightarrow{\sigma}^\sigma(\mathbf{r}, \mathbf{r}') \cdot \mathbf{E}^\sigma(\mathbf{r}') = w^\sigma(\mathbf{r}). \qquad (2.54)$$

Another condition comes from the definitions of the internal field, Eq. (2.47), and the vertex function, Eq. (2.45); then

$$\int_C \mathbf{E}^\sigma(\mathbf{r}') \cdot d\mathbf{r}' = V \qquad (2.55)$$

where $V$ is the voltage applied to the outer boundaries of the structure and the line integral is evaluated along the current path $C$ from one boundary to the other. In effect, the condition Eq. (2.55) has to be satisfied for both channels $\sigma$.

In cases where there is some symmetry (layered structures), we will show that one can solve for the fields and currents. For more complicated geometries (granular films), we resort to an ansatz based on some degree of randomness to find the measured resistivity and MR [19].

## 2.3 Layered structures

The general constitutive relation between field and current given by Eqs. (2.48) and (2.49) simplify when there are spatial and magnetic (spin) symmetries. For a magnetic multilayer if we perform random impurity averages over the impurity potentials and do not consider magnetic domains, the symmetry of the lattice is restored, and momentum in the plane of the layers $k_\parallel$ is a good quantum number. If the electric fields are uniform across the layers only $k_\parallel = 0$ enters and Eq. (2.48) is written as [19]

$$\mathbf{j}^\sigma(z) = \int dz' \overleftrightarrow{\sigma}^\sigma(z, z') \cdot \mathbf{E}^\sigma(z'), \qquad (2.56)$$

so that we are left with a one-dimensional problem for the spatial dependence of currents and fields.

For fields parallel to the plane of the layers (CIP), $\mathbf{E}^\sigma(z')$ is a constant $E$ so that the measured current per unit area, i.e., the average current density is

$$\langle j(z) \rangle = \sum_\alpha \langle j^\sigma(z) \rangle = \frac{1}{L} \int dz j(z) \equiv \sigma_{CIP} E \tag{2.57}$$

where $\sigma_{CIP}$ is the measured CIP conductivity

$$\sigma_{CIP} = \frac{1}{L} \sum_\sigma \int \int dz dz' \sigma_\parallel^\sigma(z, z') \tag{2.58}$$

and $L$ is the length of the lead in the $z$ direction over which current is probed. The symbol $\parallel$ denotes the spatial component of the conductivity tensor in the plane of the layers.

For CPP the internal field $E^\sigma(z')$ is *not* constant; nonetheless we obtain information from the continuity equation, Eq. (2.53) on the current density. Therefore it is better to invert Eq. (2.48) and express the electric field in terms of the current

$$E^\sigma(z) = \int dz' \rho_\perp^\sigma(z, z') j^\sigma(z') \tag{2.59}$$

where $\rho_\perp$ is the component of the resistivity tensor perpendicular to the plane of the layers. As the Kubo formalism yields conductivities we find the two-point resistivity by solving the integral equation

$$\int dz'' \sigma_\perp^\sigma(z, z'') \rho_\perp^\sigma(z'', z') = \delta(z - z'). \tag{2.60}$$

From Eqs. (2.59) and (2.55), the voltage drop per unit length or average electric field is

$$\overline{E}^\sigma = \overline{E}$$
$$= \frac{1}{L} \int \int dz dz' \rho_\perp^\sigma(z, z') j^\sigma(z') \tag{2.61}$$

where $L$ is the distance between leads in the $z$ direction. The CPP current density satisfies the ordinary differential equation (see Eq. (2.53))

$$\partial_z j^\sigma(z) = w^\sigma(z). \tag{2.62}$$

In the case of no spin-flips $w^\sigma(z) = 0$ the spin channels conduct independently of one another, so that $j^\sigma(z)$ is a constant, and one finds the measured current density is [19]

$$j = \sum_\sigma j^\sigma = \overline{E} \sum_\sigma [\rho_{CPP}^\sigma]^{-1}, \tag{2.63}$$

so that the CPP conductivity, when conduction in the two spin channels is independent of one another, is

$$\sigma^{CPP}(w=0) = \sum_{\sigma} [\rho^{\sigma}_{CPP}]^{-1},  \qquad (2.64)$$

where $[\rho^{\sigma}_{CPP}]^{-1}$ is the inverse of the *averaged* two-point resistivity is

$$\rho^{\sigma}_{CPP} = \frac{1}{L} \int \int dz dz' \rho^{\sigma}_{\perp}(z, z').  \qquad (2.65)$$

By comparing the conductivities for CIP and CPP Eqs. (2.58) and (2.64) we can see they are quite different. The CIP global conductivity is a complete sum of the two-point conductivities over the entire sample and over spin indices; in some limiting cases it is analogous to conduction for a set of resistors in parallel. The CPP conductivity is considerably more convoluted; from Eqs. (2.64) and (2.65) it is a sum over the conductivities for each spin channel $\sigma$; the conductivity in each channel comes from taking the inverse of the resistivity $\rho^{\sigma}_{CPP}$ which is arrived at by summing over the two-point resistivities. As we will show in some limiting cases it is analogous to a set of resistors in series, for a specific spin index.

The form of the constitutive relation Eq. (2.56) and $\mathbf{j} = \sum_{\sigma} \mathbf{j}^{\sigma}$ constitute the two current model as it was originally introduced [11]. In homogeneous media electric fields are uniform, and the condition (2.55) requires the fields to be independent of spin. The appropriate generalization to inhomogeneous magnetic structures are more complicated as they require the use of spinor currents, see Eq. (2.43) and Ref. [19]. In general, the inversion of the two point conductivity, see Eq. (2.60), remains a formidable problem. Under certain limiting cases this does simplify as we will now discuss.

## 2.3.1 Limiting cases

It is relatively straightforward to calculate conductivity for the CIP geometry; at least without vertex corrections, i.e., Eq. (2.25) , (2.19) or Eq. (2.49). However for CPP one has to find the two point resistivity which involves solving the integral equation Eq. (2.60); therefore it is important to use approximations which are exact in limiting cases. The form of the two-point conductivity is controlled by (1) the distribution of the scatterers and (2) the intensity of the scattering. The first is dictated by the layering and is characterized by $d_{in}$, the characteristic length scale for inhomogeneities, e.g., the repeat distance for the motif in a multilayer. The second is characterized by the mean free path $\Lambda$, which represents the average of the scattering encountered over the distances $d_{in}$. There are two limits for which $\sigma(z, z')$ of Eq. (2.56) is simple: (1) when $\Lambda \gg d_{in}$ and (2) when $\Lambda \ll d_{in}$.

When $\Lambda \gg d_{in}$, the conductivity is "self-averaging" and the layering or granularity is not important. This limit resembles the case for homogeneous alloys and will be referred to as such. In this case inhomogeneities are irrelevant and translational invariance is restored; in reciprocal space, the two-point conductivity is diagonal:

$$\overleftrightarrow{\sigma}(\nu, \nu') = \overleftrightarrow{\sigma}(\nu)\delta(\nu - \nu'), \tag{2.66}$$

with $\overleftrightarrow{\sigma}(\nu)$ being the Fourier transform of the real-space conductivity kernel $\overleftrightarrow{\sigma}(z - z')$. Now the conductivity matrix is readily inverted in reciprocal space, see Eq. (2.60), so that

$$\rho_{\perp}^{\sigma}(\nu) = \frac{1}{\sigma_{\perp}^{\sigma}(\nu)}, \tag{2.67}$$

and from Eq. (2.64) the CPP conductivity is

$$\sigma_{CPP}|_{\Lambda \gg d_{in}} = \sum_{\sigma=\uparrow,\downarrow} \sigma_{\perp}^{\sigma}(\nu = 0), \tag{2.68}$$

as a divergenceless current ($j^{\sigma} = $ constant) implies that $\nu = 0$. This result is reasonable as the layering is imperceptible in this limit.

When $\Lambda \ll d_{in}$, which is called the local limit, the conductivity tensor is a one-point function,

$$\overleftrightarrow{\sigma}(z, z') = \overleftrightarrow{\sigma}(z)\delta(z - z'), \tag{2.69}$$

for points $z$ and $z'$ separated by distances $|z - z'| \gg \Lambda$. In this limit the conductivity is a diagonal matrix, which is easily inverted, so that

$$\rho_{\perp}^{\sigma}(z, z') = \frac{1}{\sigma_{\perp}^{\sigma}(z)}\delta(z - z'). \tag{2.70}$$

By placing this in Eq. (2.59) and for no spin-flips ($w = 0$), we find the remarkably simple result that the spin-dependent internal fields are proportional to the inverse of the conductivity

$$E^{\sigma}(z) = \frac{j^{\sigma}}{\sigma_{\perp}^{\sigma}(z)}, \tag{2.71}$$

where $j^{\sigma}$ is the constant current density in the $\sigma$ conduction channel. Also when we use Eq. (2.70) in Eqs. (2.65) and (2.64) we find

$$\sigma_{CPP}(w = 0)|_{\Lambda \ll d_{in}} = \sum_{\sigma=\uparrow,\downarrow} \left[\frac{1}{L}\int_0^L \frac{dz}{\sigma_{\perp}^{\sigma}(z)}\right]^{-1}, \tag{2.72}$$

which is quite different from the CIP conductivity which is found by placing Eq. (2.69) in Eq. (2.58)

$$\sigma_{CIP}\big|_{\Lambda \ll d_{in}} = \sum_{\sigma=\uparrow,\downarrow} \frac{1}{L} \int_0^L dz \sigma_\parallel^\sigma(z) = \frac{1}{L} \int_0^L dz \sum_{\sigma=\uparrow,\downarrow} \sigma_\parallel^\sigma(z). \qquad (2.73)$$

In the CIP geometry and with the local character of the conductivity (Eq. (2.69)), we find the global conductivity (Eq. (2.73)) is independent of the orientation of the magnetization of one layer relative to another. Note that $\sum_\sigma \sigma^\sigma(z)$ enters before integration. This is not the case for CPP, where, in Eq. (2.72), the sum over spin is taken only after one has evaluated the global conductivity for each spin channel; the latter is sensitive to average magnetization in the layered structure. From this we conclude that no MR exists in the local limit for CIP; however, it does exist for CPP with $w = 0$. Also, in the local limit it is meaningful to talk about the conductivity or resistivity of each layer independent of the neighboring layers, and it is appropriate to make analogies with resistor networks to understand whether or not the resistivity changes in going from an antiferromagnetic or random configuration to a ferromagnetic alignment of the magnetic layers.

In Fig. 2.4(a), we show the appropriate network corresponding to four layers - two magnetic and two non-magnetic - in CIP (Eq. (2.73)). We see that the resistivities are the same for the P and AP configurations, so that there is no CIP-MR in this limit. For CPP ($w = 0$), where there is no mixing of the currents (for electrons with spin-up with those of spin-down), the resistances of the individual layers for each spin direction are added in series, while those for the two channels are added in parallel. In Fig. 2.4(b), we note that in the ferromagnetic configuration the resistance is less in one branch than in the other, producing a "short-circuit" effect, while in the antiferromagnetic configuration they are equal. Thus, even in the local limit we have MR for CPP ($w = 0$). When the currents are mixed via spin-flips, $w \neq 0$, the resistors are coupled and the effect of the short circuit is reduced so that the CPP-MR is diminished.

## 2.3.2 Scattering encountered

As we saw above the expression for the conductivity of a multilayer is quite different for currents parallel and perpendicular to the plane of the layers. There are two reasons: first, the component of the conductivity tensor parallel to the layers $\sigma_\parallel(z, z')$ is different from that perpendicular $\sigma_\perp(z, z')$; this is related to the electronic structure of the multilayer and will be elaborated on in Sec. 2.5. The second is the scattering encountered by electrons as they traverse the structure is different.

For currents in the plane of the layers transport is sensitive to all the scattering in an atomic plane as the field drags the electrons in this direction; however in the direction normal to the current, and therefore perpendicular to

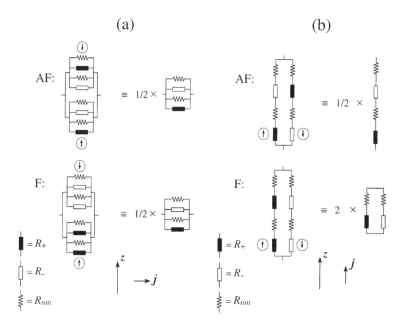

Figure 2.4: Resistor network analogy for the (a) CIP and (b) CPP resistances of the ferromagnetic (P) and antiferromagnetic (AP) configurations of the magnetic moments in a multilayer in the local limit $\lambda \ll d_{in}$. $R_+, R_-$, and $R_{nm}$ stand for the resistances of the magnetic layers for spin parallel and antiparallel to the local magnetization, and of the non-magentic layers. In CIP (a) the current density $\mathbf{j}$ is perpendicular to the growth direction $z$, while for CPP (b) it is parallel to $z$; taken from [19].

the plane of the layers, transport is sensitive only to the scattering in a region limited by the mean free path about the plane for which one is evaluating the current. An electron retains memory of its momentum only within this distance.

In Fig. 2.5(a), we show the cylindrical region which influences transport along its' axis; all the scattering in this region must be included to determine the current at the center. This has the consequence that if the impurity scattering is sufficiently strong (either due to concentration of impurities or strength of individual scattering) so that the mean free path is short or the distance between adjacent magnetic layers is long the cylindrical region which influences transport about a plane never encompasses two magnetic layers. In this limit, the reorientation of the moments of adjacent magnetic layers by an external field does not influence the transport in this geometry, i.e., there is no

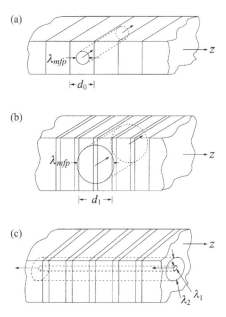

Figure 2.5: The volume of scatterers sampled by electrons for CIP (a,b) and CPP (c); it is only the scattering within these regions that control the current at the center of the tube. While electrons travel in all directions the net current is parallel to the electric field in the sample; therefore the cylindrical tubes represent the motion the component of their motion along the direction of the field. The radius of the tube is determined by the mean free path $\lambda_{mfp}$. (a) The scattering is sufficiently strong so that $\lambda_{mfp}$ is smaller than the thickness of a layer. In this "local" limit the current does not sample the scattering from different layers; in particular it is insensitive to the scattering in the adjacent magnetic layer, so that there is no CIP-MR. Also the local resistivity $\rho(z)$ is strongly position dependent. (b) The $\lambda_{mfp}$ is large compared to the spacing between layers so that the current is influenced by scattering in different magnetic layers and one has a CIP-MR; in this "homogeneous" or self averaging limit the resistivity is independent of position $z$. (c) For the CPP configuration the current samples scattering in all layers. In this case the CPP-MR does not depend on $\lambda_{mfp}$; however it will depend on the spin diffusion length which we have not considered in this discussion.

CIP-MR. Therefore, in the CIP geometry the mean free path of the electrons is a relevant length scale for MR. Parenthetically, the current measured by leads covering a set of atomic planes add the currents in each of the atomic plane in parallel; this has the consequence that broad leads measure an average of the current in the individual atomic planes. This does not negate the observation that the CIP-MR vanishes when the mean free paths are not long enough to cover adjacent magnetic layers.

When current is perpendicular to the plane of the layers transport is sensitive to scattering in a cylinder whose axis is normal to the layers; see Fig. 2.5(b). Now the direction of the current guarantees that transport samples the scattering in different layers quite independently of the mean free path. This raises the question what limits the scattering that influences transport along the current direction. We know that the inelastic mean free path or distance over which the energy eigenstates remains coherent is an upper limit for all transport, however the spin direction of the electron is another label for eigenstates, and Valet and Fert [21] defined a new length scale, the spin diffusion length $\Lambda_{sdl}$, over which the electron's spin remains in an eigenstate. At low temperatures, which will be the primary focus in this chapter, there are many more scattering events that flip the spin's direction than alter its energy, i.e., $\Lambda_{sdl} \ll$ the coherence length. Furthermore, there are many more elastic collisions that merely alter the electrons momentum direction than flip its' spin in most of the GMR devices studied to date so that $\Lambda_{mfp} \ll \Lambda_{sdl}$. As $\Lambda_{mfp}$ is not a relevant length scale for CPP transport the next larger length $\Lambda_{sdl}$ is the controlling one. In the plane of the layers, the distribution of impurities is assumed uniform so that currents along all axes normal to the planes are the same. In the presence of magnetic domain formation in the plane of the magnetic layers the mean free path could become a relevant parameter, however this is usually not the case because their size is large in comparison to the mean free path in GMR devices.

### 2.3.3 Approaches

Superficially the three formalisms for calculating transport properties seem quite different. The Boltzmann approach describes transport in terms of non-equilibrium distribution functions, the Landauer formalism equates transport to quantum mechanical transmission amplitudes and probabilities across the structure, and the Kubo formula gives the current in terms of equilibrium correlation functions. Nonetheless they have been shown to be related to one another; for the Kubo and Landauer approaches by Fisher and Lee [22], and by Mahan amongst others for the Kubo and Boltzmann formalisms [17]. While it is difficult to see how one can apply the Landauer approach other than to the entire structure the other two can be solved piecewise as we now discuss.

### 2.3.3.1    Layer by layer

At some level one can look upon a magnetic multilayer as an assembly of alternating magnetic and non-magnetic layers, and solve the Boltzmann equation for each region separately. To relate distribution functions across interfaces one specifies boundary conditions that provide for continuity of current. This was done for CIP geometries, first where the diffuse scattering from impurities and roughness at interfaces was modeled in terms of transmission and reflection coefficients [23], and later where specular scattering at interfaces which gives rise to (specular) reflection and transmission coefficients for electrons as they traverse regions with varying potentials, was included as well as diffuse scattering [24]. This leads to relations of the distribution functions on the two sides (1 and 2) of an interface, such as

$$f(\mathbf{v}_2^>) = R(\mathbf{v}_1, \mathbf{v}_2)S(\mathbf{v}_1, \mathbf{v}_2)f(\mathbf{v}_2^<) + T(\mathbf{v}_1, \mathbf{v}_2)S(\mathbf{v}_1, \mathbf{v}_2)f(\mathbf{v}_1^>) \qquad (2.74)$$

and

$$f(\mathbf{v}_1^<) = R(\mathbf{v}_1, \mathbf{v}_2)S(\mathbf{v}_1, \mathbf{v}_2)f(\mathbf{v}_1^>) + T(\mathbf{v}_1, \mathbf{v}_2)S(\mathbf{v}_1, \mathbf{v}_2)f(\mathbf{v}_2^<) \qquad (2.75)$$

where $R$ and $T$ are the reflection and transmission coefficients and $1 - S$ represents the diffusive scattering, i.e., when $S = 1$ there is none. The arrows on top of the velocities indicate whether the distribution function is for electrons moving normal to the interface with $v_z \lessgtr 0$. The first equation relates the distribution function to electrons moving away from an interface in terms of the function for electrons approaching the interface from the same side and being reflected $R(\mathbf{v}_1, \mathbf{v}_2)$, and those being transmitted across $T(\mathbf{v}_1, \mathbf{v}_2)$ the interface. For CPP there is current across the interface, and non-equilibrium accumulation of charge and spin [25]. To guarantee continuity of current under the general conditions of diffuse as well as specular scattering it is necessary to add an additional term to the right hand sides of Eqs. (2.74) and (2.75) of the form [26]

$$[1 - S(\mathbf{v}_1, \mathbf{v}_2)] \frac{1}{\Omega} \int d\mathbf{k}' \left[ |v_{2z}'| f(\mathbf{v}_2'^<) + |v_{1z}'| f(\mathbf{v}_1'^>) \right]. \qquad (2.76)$$

The distribution function for each individual layer is found from the Boltzmann equation usually by making the assumption that the scattering rate or relaxation time is constant within a layer. In this case, the piecewise solutions are straightforward, and the boundary conditions Eqs. (2.74) and (2.75) connect them to find the distribution function for the entire multilayer. This approach is reasonable when the thicknesses of the layers is large compared to the mean free paths, $\Lambda \ll d_{in}$, because transport in a layer will not be affected by the scattering in adjacent layers. In this *local limit* each layer is viewed as a resistor with properties independent of the neighboring layers, and analogies to resistor networks to understanding electrical transport in multilayers are useful. However, when $\Lambda \gg d_{in}$ the assumption of a constant scattering rate across a layer breaks down; while there are ways of incorporating these variations they lead to complications and make it at least as cumbersome as solving for the distribution function across the entire structure.

### 2.3.4   Whole structure

In the *homogeneous* and especially *ballistic limits* transport in multilayers is controlled by the electronic band structure of the entire system. On the average electrons travel across several layers before scattering so that it is important to have wavefunctions that extend over these region. It is not difficult to obtain the band structure for multilayered structures by the methods we describe below, but the spatial variations of the diffuse scattering from impurities in the bulk and interface roughness, see Eq. (2.9), complicates the solution of the transport equations. Analytic solutions have been obtained for a free electron model, i.e., $V_{pot,M}^{\sigma}(\mathbf{r}) = 0$, see Eq. (2.6), by using the Kubo formalism in momentum space by making some plausible but unproven approximations [27]. These solutions have given us insight into how the conductivity at one point is affected by the scattering in the region circumscribed by the mean free path; in particular it has confirmed the exponential decay of the CIP-MR as the distance between magnetic layers increases relative to the average mean free path in the region. The Boltzmann equation can be applied to the entire multilayer by using realistic band structure [14, 28, 29, 30]; this has yielded useful information in the dilute limit.

To simultaneously account for effects of band structure and reasonable densities of impurities it has been necessary to resort to numerical calculations. One easily obtains the electron band structure for multilayers by using layered versions of the ab-initio methods available for homogeneous materials, e.g., the Korringa-Kohn-Rostoker (KKR) method [32, 33]. However obtaining ab-initio results for the conductivity is quite time consuming. This has been attempted in all three approaches (Boltzmann, Landauer and Kubo) and will be described below in the section on *Origin of GMR*. What we have learned to date is that it is essential to do these types of calculations as the approximations made in simpler approaches have forced one to use unrealistic values for the parameters describing the scattering due to impurities. In particular the self-consistent determination of the scattering and its effect on the band structure seems to be crucial for obtaining agreement with experimental data on the systematic changes in conductivities as we vary the material parameters, i.e., type of metallic layers used, impurity distribution (in the bulk of the layers or at interfaces) and their concentration. Underlying these methods to date is the single-site coherent potential approximation (CPA), and that vertex correction have not been included [34, 35].

## 2.4 Two current model

### 2.4.1 Applicability

The current density in solids is a sum of conduction over eigenstates of the Hamiltonian for the electronic structure Eq. (2.6). As we are only considering non-relativistic cases, the labels consist of spin $\sigma$ and state $k$. For materials with collinear magnetic structure spin is a good quantum number, as long as we choose the quantization axis that makes the potential $V^\sigma_{pot,M}(\mathbf{r})$ in Eq. (2.6) diagonal. For these cases we gather all the conduction paths into two groups $\alpha =\uparrow,\downarrow$, and thereby arrive at the two current model of electron conduction in magnetic materials [11]. The potential that scatters electrons in magnetic metals is spin dependent, see Eq. (2.9); there are spin-flip processes and non spin-flip ones (see also Eq. (2.21)). The latter are diagonal when the intrinsic potential $V^\sigma_{pot,M}(\mathbf{r})$ is diagonal [27]; this leads to spin dependent scattering events that change the momentum $k$ of the electrons without altering their spin. When we exclude spin-flip scattering processes and make the gross yet conventional assumption that all $k$ states for a spin direction scatter at the same rate we arrive at a particularly simple parameterization of conduction in which one assigns a scattering rate or resistivity to each spin channel of conduction (see also Eq. (2.30)):

$$\rho^{\uparrow,\downarrow} = \rho^{M,m} \equiv a \pm b, \tag{2.77}$$

where the superscripts $M$ and $m$, refer to electrons with spin parallel (Majority) and opposite (minority) to the magnetization. From Eq. (2.65) and the realization that for a homogeneous sample the effective fields are independent of spin, the total current in the two spin channels is,

$$j = j^\uparrow + j^\downarrow = (\sigma^\uparrow + \sigma^\downarrow)E, \tag{2.78}$$

so that the resistivity is

$$\rho = \frac{1}{\sigma^\uparrow + \sigma^\downarrow} = \frac{\rho^\uparrow \rho^\downarrow}{\rho^\uparrow + \rho^\downarrow} \tag{2.79}$$
$$= \frac{a^2 - b^2}{2a}.$$

It is obvious that the resistance in ferromagnetic metals is less than in materials with comparable scattering rates, i.e., the average resistivity for each channel is $a$, so that for two channels conducting in parallel we find $\rho = 1/2a$, while from Eq. (2.79) we find $\rho < \frac{1}{2a}$. The channel with the lower resistivity conducts more of (shunts) the current, and creates the effect of a *short circuit*.

## 2.4.2   Consequences

This line of reasoning provides an immediate explanation for the GMR produced by the spin dependent scattering in magnetic multilayers [36]. We consider a multilayer made of alternating magnetic and non-magnetic layers, e.g., Co/Cu or Fe/Cr, of equal thicknesses, periodically repeated as in a superlattice. As the magnetic layers can have their magnetizations point parallel or antiparallel to one another the repeat motif consists of four layers; 2 magnetic, 2 non-magnetic. We will adopt the homogeneous or self-averaging limit, see Sec. 2.3.1, so that the resistivity in each channel is an *average* of the scattering encountered over the 4 layers. As we do not consider spin-flips the resistivity for each spin channel for CIP is

$$\rho^\uparrow = \frac{1}{4}\left[2\rho^M + 2\rho^n\right], \qquad (2.80)$$

$$\rho^\downarrow = \frac{1}{4}\left[2\rho^m + 2\rho^n\right],$$

so that from Eq. (2.79) the resistivity in the parallel or ferromagnetic configuration of magnetic layers is

$$\rho_P = \frac{1}{4}\left\{(a+c) - \frac{b^2}{(a+c)}\right\}, \qquad (2.81)$$

where $\rho_n = c$ is the resistivity for the non-magnetic layer. The up spin scatters as a Majority electron in the magnetic layers while the down as a minority electron. For the antiferromagnetic alignment of the layers the resistivity for the spin channels are

$$\rho^\uparrow = \rho^\downarrow = \frac{1}{4}\left[\rho^M + \rho^m + 2\rho^n\right], \qquad (2.82)$$

as an electron is scattered as a majority electron in the layer where its spin is parallel to the magnetization, and as a minority electron in the layer where it is opposite. From Eq. (2.79) the resistivity for the antiferromagnetic configuration is

$$\rho_{AP} = \frac{1}{4}(a+c). \qquad (2.83)$$

By comparing the resistivities Eqs. (2.81) and (2.83) we find $\rho_P < \rho_{AP}$ and there is a negative MR when an antiferromagnetically aligned multilayer is aligned ferromagnetically; this is called the GMR effect. In this picture, the role of the applied magnetic field is solely to realign the magnetic layers; the action of the fields on the electrons, the Lorentz force, is not accounted for. The origin of this GMR effect is the introduction of a short circuit when the layers are ferromagnetically aligned [36].

This explanation focused on the role of spin dependence of the scattering, Eq. (2.9), and does not address the effect of the spin dependent potentials, Eq. (2.6) in producing GMR. In the derivations above we have assumed that the wavefunctions (bandstructure) for the ferro and antiferromagnetic configurations of the magnetic multilayer are the same; only the scattering of the electrons changes. While the above toy model is useful to understand some aspects of GMR, as we discuss in Sec. 2.5.2, the band structure does change and it is important to take this into account. Also in the above we placed ourselves in the homogeneous limit where the MR from solely scattering is the same for CIP and CPP. In this limit, the resistor network analogy we used is tenuous for the reasons outlined in Sec. 2.3.3.1, therefore one should not take the above toy model explanation of GMR too literally. In the local limit where the resistor network is more applicable CIP-MR vanishes, while the CPP-MR is still given by the difference between the resistivities Eqs. (2.81) and (2.83), and the above analysis is more credible (see Eq. (2.72)) to see how to add resistivities of the individual layers for CPP.

While we focused on scattering in the bulk of the layers, it is possible to extend the above analysis to include interfaces. As the approximations made are more tenuous for interfacial scattering and as it does not further enlighten us as to the two current model, we do not consider it here.

## 2.4.3 Breakdown

In the above discussion we neglected spin-flip scattering; when one includes it the electron's spin direction is no longer conserved so that the two spin currents are mixed, i.e., they are not independent of one another. At low temperature the primary source of spin-flip scattering comes from impurities with spin-orbit coupling or paramagnetic ones for which spin-flip is an elastic process. As most of the scattering encountered in magnetic multilayers at low temperature does not flip spin as it costs energy, i.e., it is inelastic, the spin diffusion length is usually much longer than the mean free path [21], so that for CIP where transport is controlled by $\Lambda_{mfp}$ the presence of spin-flip processes and the subsequent breakdown of the two current model is not perceived. However for CPP, where $\Lambda_{mfp}$ does not control GMR, spin-flips limit the distance over which the two spin currents are independent, and concomitantly reduce the MR in this geometry. This has been nicely demonstrated in a series of experiments in which the amount of spin-flip scatterers, spin-orbit coupled and paramagnetic impurities, has been shown to mix the two currents and significantly reduce the GMR [37]. At higher temperatures inelastic spin-flip processes occur without introducing impurities, e.g., electron-magnon scattering, however as stressed by Fert, they do not increase the resistivity assigned to each channel, because momentum is conserved [11].

The basis for the simple parameterization of transport in the two current model is assigning one scattering rate to all the currents with one spin direction. While this may have some validity for homogeneous materials it is not correct for multilayers, except for the local limit which is attained in magnetically layered nanowires [38]. As we have seen in Sec. 2.2, scattering depends on momentum as well as spin; there are as many scattering rates as there are states in a multilayered structure. Parenthetically, as transport is primarily confined to the Fermi surface, by momentum we mean the Fermi momentum. Therefore it is important to caution that the two current model as used to understand transport in ferromagnetic alloys is not really applicable to magnetic multilayers. While it is not impossible for toy models to account for this, ab-initio calculations are ideally suited for this task [12, 39]. Calculations which are completely ab-initio in no way invoke the two current model; however these calculations have been found to be extremely time consuming. One compromise has been to do ab-initio calculations of the band structure, and assignment of different spin-dependent scattering rates to layers and interfaces that are independent of the electron's momentum [40]. This provides a convenient way to include the effects of band structure and a parameterization of the scattering in the different regions with a reasonable number of unknown constants that are determined by fits to data. The concern one might have is that the neglect of the dependence of scattering rates on momentum yields unrealistic scattering rates.

One final source for breakdown of the independence of the up and down spin currents is the non-collinearity of the spin structure. For non-collinear structures the eigenstates are not pure spin states, so that when electrons undergo non spin-flip scattering they nonetheless mix the currents being conducted in eigenstates. This mixing has been identified as a source resistivity of walls between oppositely oriented magnetic domains [41].

## 2.5   Origins of GMR

Magnetoresistance has several origins; the most common one in metals, the effect of magnetic fields on the trajectory of the electrons (the Lorentz force), does not contribute much to metallic layered films because of their high residual resistivity. The frequent scattering of an electron does not allow the magnetic field to act long enough to curve its trajectory. As a multidomain sample is magnetized, domain walls are erased. This effect contributes to the MR observed in pure iron whiskers; however, the change in resistivity produced by this is dwarfed [42] by the resistivity of the multilayered structures themselves, so that this cannot be the origin of their GMR ratios. Anisotropic magnetoresistance (AMR), i.e., the variation of the resistivity with the angle between the current and magnetic field, is the origin of the MR in the permalloy films that are used in a former generation of magnetoresistive reading heads and can produce an MR ratio of several percent; it is an order of magnitude too small to explain the GMR observed in multilayered structures. In addition, the MR observed in multilayers and granular films is almost entirely isotropic; therefore, AMR cannot be its origin.

All layered and granular structures that display GMR are made of at least two different metals; one of them magnetic. Thus, the electronic structure changes from one metal to another. If the length scale associated with this inhomogeneity ($d_{in}$) is less than the mean free path ($\Lambda_{mfp}$) due to other scattering, the different metals become a source of scattering, e.g., for granular films with small ($\sim 15\,\text{Å}$) magnetic precipitates. However, superlattices are periodic, and changes in band structure are not a source of diffusive scattering; rather the specular scattering due to periodic compositional differences change the dispersion and the wave functions. The conduction bands for composites are different from those of the constituents. These differences affect the scattering of conduction electrons by other sources. In layered and granular structures, the external field reorients one magnetic region relative to another and thereby alters the composite band structure. In driving a system from antiferromagnetic to ferromagnetic, one changes the band structure from one in which the majority and minority spin bands alternate to one in which they remain constant as one goes from one region to another. The dependence of the wave functions on the magnetic configuration of the composite structure changes the conduction electron scattering rates, even when one evaluates a scattering event that by itself is independent of spin. Compositional differences in these structures, although they may not be a source of scattering, have a profound effect on how scattering events are felt by the conduction electrons through their wave functions and thereby are an essential ingredient for the GMR observed in magnetically layered structures.

As each electron propagator has its own rate of decay only ab-initio calculations of the magnetotransport properties of magnetic multilayers are able to capture this and do a credible calculation of GMR with two major caveats: that one knows the impurities-defects and their distribution that give rise to scattering, and that one is able to faithfully model the scattering they produce. In the following subsection we review the progress to date in this direction; however we will first review the conclusions and *analytic* results one is able to make based on the theory outlined above.

## 2.5.1    Analytic results

There has been a reasonable number of analytic results [27]; here we single out one that uses the free electron model and focuses on the inhomogeneity of the scattering due to impurities and interfacial roughness [43]. The one point conductivity for layered structures is calculated in momentum space by neglecting vertex corrections, see Fig. 2.3, i.e., we calculated the bubble conductivity Eq. (2.50) adapted for layered structures and integrated over $z'$. In momentum space the self energy of electrons is nonlocal and we made a plausible decoupling approximation to arrive at the CIP conductivity, Eq. (2.58)

$$\sigma_{CIP} = \frac{ne^2}{2m} \sum_\sigma \frac{1}{L} \int \frac{dz}{a^\sigma(z)}, \tag{2.84}$$

where

$$a^\sigma(z) = \frac{1}{\Lambda^\sigma} \sum_a \Delta_a^{\sigma\sigma} e^{-|z-z_a|/\Lambda^\sigma}. \tag{2.85}$$

The sum over $a$ is over all scattering centers and $\Delta_a^{\sigma\sigma}$ represents the scattering rate at $a$

$$\Delta_a^{\sigma\sigma} = \rho(E_F) \langle v_a^2 \rangle \left(1 + p_a^2 \pm 2p_a\right), \tag{2.86}$$

where $\pm$ refers to the spin of the electron $\sigma$ being up or down, $p \equiv j/v$, and $v_a, j_a$ are the parameters characterizing the scattering potential, see Eq. (2.9). The mean free path $\Lambda^\sigma$ is derived from the average scattering in the entire structure

$$\Lambda^\sigma = \frac{k_F/m}{\bar{\Delta}^\sigma}, \tag{2.87}$$

where

$$\bar{\Delta}^\sigma = \frac{1}{L} \sum_a \Delta_a^{\sigma\sigma}. \tag{2.88}$$

To obtain the CPP resistivity one needs the two point resistivity defined by Eq. (2.60); an approximation (albeit somewhat uncontrolled, it is correct in the local limit, see Eq. (2.70)) is to take the inverse of the one-point conductivity used to find the CIP conductivity Eq. (2.84)

$$\rho^\sigma(z) = \frac{2m}{ne^2} a^\sigma(z). \tag{2.89}$$

By averaging over the length of a sample we find

$$\rho^{\sigma}_{CPP} = \frac{2m}{ne^2} \frac{1}{L} \int_0^L a^{\sigma}(z) dz$$

$$= \frac{2m}{ne^2} \bar{\Delta}^{\sigma}. \tag{2.90}$$

The resistivity in each spin channel is given by the total scattering in that channel; this is called *self-averaging*. By placing this result in Eq. (2.72) we find the CPP conductivity is

$$\sigma_{CPP}(w = 0) = \frac{ne^2}{2m} \sum_{\sigma} \frac{1}{\bar{\Delta}^{\sigma}}, \tag{2.91}$$

which is an example of the two current model as it applies to a magnetic multilayer. It should be noted we neglected spin-flip processes to arrive at this result.

The results obtained, Eqs. (2.84) and (2.91), illustrate the main features of the transport in the CIP and CPP geometries. In the first current at one point $z$ is controlled by $a^{\sigma}(z)$ which represents an average of the scattering in the neighborhood of $z$ weighted by its distance from the scatterers relative to the mean free path $\Lambda^{\sigma}$, see Eq. (2.85). Clearly if the thickness of the non-magnetic layers are larger than $\Lambda^{\sigma}$ transport in one magnetic layer is unaffected by the scattering of electrons in an adjacent layer and there is no scattering contribution to the GMR effect. In this case which is called the *local limit* there can still be a contribution to GMR from changes in the electronic structure as the multilayer goes from antiferro to ferromagnetic alignment, however this is not accounted for in a free electron model where all layers have the same potential $V^{\sigma}_{pot,M}(\mathbf{r}) = const$, see Eq. (2.6). On the contrary, CPP transport is not controlled by the weighted average of scattering in a region limited by the $\Lambda^{\sigma}$, see Eq. (2.90), so that the GMR effect survives as long as the spin diffusion length $\Lambda_{sdl}$ is longer than the distance between magnetic layers.

For a bipartite superlattice, e.g., Co/Cu, in the homogeneous limit where Eqs. (2.84) and (2.91) yield the same result, the bounded MR ratio, Eq. (2.2) is [44]

$$R' = 4 \left[ \frac{m p_m w_m + 2 p_s w_s}{m(1 + p_m^2) w_m + n w_{nm} + 2(1 + p_s^2) w_s} \right]^2, \tag{2.92}$$

where $m$ and $n$ are the number of MLs in the magnetic $m$ and non-magnetic $nm$ layers, $p_{m/s}$ are the ratios, see Eq. (2.86), giving the spin dependence of the scattering (the index $s$ denotes the interface scattering), and $w_i \equiv \rho_i(E_F) \langle v_i^2 \rangle$ where $i = m, nm, s$, see Eq. (2.86). In the absence of spin-dependent scattering $p_{m/s} = 0$ there is no GMR. To obtain the largest $R'$ one should minimize the thickness of the non-magnetic layers $b$. The maximum value $p_i = \pm 1$ yields an MR ratio of

$$R'(p = \pm 1) = \left[ 1 + \frac{\frac{1}{2}nw_{nm}}{mw_m + 2w_s} \right]^{-2}. \tag{2.93}$$

As we reduce either $n \longrightarrow 0$ or $w_{nm} \longrightarrow 0$, $R' = 1$; however there are limits as to how much one can reduce $n$ due to the coupling between magnetic layers across the non-magnetic spacer. By using the resistivities Eqs. (2.81) and (2.83) from the toy two current model, we find

$$R' = \frac{b^2}{(a+c)^2}. \tag{2.94}$$

This reinforces the results we obtained above: we increase GMR as we reduce non-magnetic scattering, i.e., as $c \longrightarrow 0$ we increase $R'$, and as $b < a$ (from Eq. (2.77) and as $\rho \geq 0$ we know $a > b$) we see that $R' < 1$ but approaches 1 as the spin-dependent scattering increases ($b \longrightarrow a$).

With the same assumptions as above the angular dependence of the resistivity in the homogeneous limit of a bipartite superlattice that comes solely from scattering is [44]

$$\rho(H) = \frac{m}{ne^2} \frac{\gamma^2 - \beta^2 \cos^2 \theta(H)}{\gamma}, \tag{2.95}$$

where

$$\gamma = \frac{1}{l_m + l_{nm}} \left[ m(1 + p_m^2)w_m + nw_{nm} + 2(1 + p_s^2)w_s \right], \tag{2.96}$$

$$\beta = \frac{2}{l_m + l_{nm}} \left[ mp_m w_m + 2p_s w_s \right], \tag{2.97}$$

$\theta(H)$ is *one-half* of the angle between the magnetizations of adjacent magnetic layers, $l_{m/nm}$ is the thickness of the $m/nm$ layer, and we assumed the interfacial magnetization is aligned with that in the bulk of a layer. Note that in the homogeneous limit and where we consider only the effects of spin-dependent scattering the resistivities for CIP and CPP are the same. For $\theta = 0, \pi/2$, this reduces to the resistivities $\rho_P, \rho_{AP}$ that were used to derive $R'$, Eq. (2.92). In the absence of anisotropy one finds $\cos \theta$ is proportional to $H$, so that $\rho(H) \sim A - BH^2$; from existing data this is by and large not observed so that one obtains $\theta(H)$ from the relation $M(H) = 2M_o \cos \theta(H)$, when one models the magnetic layers as uniformly and rigidly ordered for all $\theta(H)$. While one expects that the electronic structure, which has been ignored in this section, affects the angular dependence of the MR this has not been seen for CIP, i.e., $\rho(H)$ depends on $\theta(H)$ as predicted by Eq. (2.95); however for CPP there are indeed deviations from the simple relation Eq. (2.95) [45].

These simple treatments of the magnetotransport in magnetic multilayers reduce it to a minimum number of parameters. While in no way adequate, for the reason given above, there are 5 parameters: the $v_a$ and $j_a$ for the spin-dependent scattering in the bulk of the magnetic layers and at interfaces, and $v_{nm}$ for the non-magnetic layer; if there is spin-flip scattering additional parameters are needed. These are treated as phenomenological parameters that are determined by fitting theoretical expressions for the resistivity and MR to data. The interpretation of these parameters, their validity and ability to explain other data, has been tested primarily for the CPP geometry; they have worked surprisingly well [46]. For CPP, the existence of MR depends on the size of the spin diffusion length relative to the inhomogeneities; as long as $\Lambda_{sdl} \gg d_{in}$ the spin currents in the two current models are independent. This seems to be the case for a large number of the magnetic multilayers studied to date; albeit there are some notable exceptions [47]. For CIP, it has been harder to extract unambiguous parameters; the expressions for the resistivity are sensitive to $\Lambda_{mfp}$ for CIP, and band structure affects transport as we discuss in the next section.

## 2.5.2 Spin-dependent potentials

### 2.5.2.1 Bragg and Fermi surface effects

All layered systems that display GMR are made of at least two different metals that are combined layer by layer: one ferromagnetic metal and the other non-magnetic. Thus, the potentials and the electronic structure change from one metal to another which is illustrated by means of the spin-projected densities of states in Fig. 2.6.

For concreteness but without loss of generality we focus on a Co/Cu multilayered structure. The electronic structure of the majority bands of Co and Cu are very similar. The majority $d$ bands are fully occupied and lie below the Fermi level. However, since Co has two less electrons than Cu, the minority

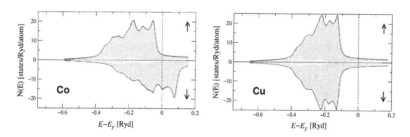

Figure 2.6: Spin-projected densities of states for Co and Cu; the differently shaded grey areas indicate the amount of $s$, $p$ and $d$ states.

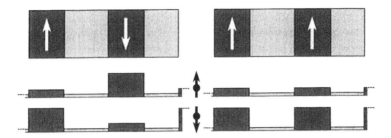

Figure 2.7: Spin-dependent potentials in a magnetic multilayer for the AP and P configurations for majority and minority electrons.

band for Co is less occupied and the electronic structure of the minority bands at the Fermi energy is very different for both metals. The majority electrons of a multilayer with their moments aligned in parallel traverse easily through the system; the reason is that the potential landscape is very flat (see Fig. 2.7) since the bandwidth is nearly the same in each layer. On the contrary minority electrons experience high potential steps (see Fig. 2.7) and are reflected at the Co/Cu interfaces. As a result, the Fermi velocity of the majority electrons is much larger than the Fermi velocity of the minority electrons, $v_F^\uparrow \gg v_F^\downarrow$ (see also Fig. 2.10). A multilayer with antiparallel moments consists of a potential with a potential well in alternate layers since the spin character of the electrons (majority, minority) changes in every other layer (see Fig. 2.7); also both spin channels are degenerate. The velocity of the electrons is mainly determined by the largest potential step. For this reason, the Fermi velocity of the electrons in an AP-ordered multilayer is nearly the same as for the minority electrons, $v_F^\uparrow \gg v_F^{AP} \geq v_F^\downarrow$ (see also Fig. 2.10).

The GMR ratio (Eq. (2.1)) calculated within the spin-independent relaxation time approximation (Eq. (2.28)) becomes [28]

$$ R = \frac{\sum_k \delta(E_k^\uparrow - E_F)\, v_i^{\uparrow 2} + \sum_k \delta(E_k^\downarrow - E_F)\, v_i^{\downarrow 2}}{2\sum_k \delta(E_k^{AP} - E_F)\, v_i^{AP2}} - 1, \qquad (2.98) $$

where $v_i$ with $i = (\|, \perp)$ are the Cartesian components of the velocity parallel and perpendicular to the layers. As we have not considered the spin dependence of the relaxation times they cancel out, and the GMR ratio is only determined by the electronic structure of the system. A detailed analysis of the GMR ratio depending on the spacer thickness (Eq. (2.98)) for $Co_n Cu_m$-multilayers is shown in Fig. 2.8. The CPP-MR is always larger than the CIP-MR, and the ratios decrease with increasing Cu-layer thickness (see also [27]). For a better understanding of these results we can approximate the GMR ratio by using the density of states at the Fermi level $N^\sigma(E_F) = \sum_k \delta(E_k^\sigma - E_F)$,

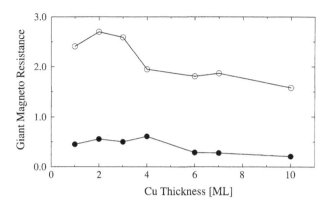

Figure 2.8: Calculated MR ratios for Co/Cu (100) multilayers as a function of the Cu layer thickness [28]. CIP-MR (filled circles) and CPP-MR (open circles).

and a Fermi surface average of the square of the velocity components

$$\overline{v_i^{\sigma 2}} = \frac{\sum_k \delta(E_k^\sigma - E_F)\, v_i^{\sigma 2}}{\sum_k \delta(E_k^\sigma - E_F)}; \qquad (2.99)$$

this leads to

$$R \approx \frac{N^\uparrow(E_F)\, \overline{v_i^{\uparrow 2}} + N^\downarrow(E_F)\, \overline{v_i^{\downarrow 2}}}{2N^{AP}(E_F)\, \overline{v_i^{AP2}}} - 1. \qquad (2.100)$$

The densities of states at the Fermi level for P and AP configurations are shown in Fig. 2.9. When one adds the majority and minority density of states in P configuration they are quite close to the density of states in AP configuration, so that they are neither responsible for the positive GMR ratios nor account for the differences between the CIP- and CPP-MR; rather these result from differences of the averaged velocities as a function of the magnetic configuration (Fig. 2.10). If we compare the averaged velocities of the majority and minority electrons in the parallel configuration with those in the antiparallel configuration, we notice that this would lead to a positive GMR ratio. For the components in the plane of the layers $v_\parallel$ the CIP-MR is obtained, whereas $v_\perp$ leads to CPP-MR. When we compare the in-plane components of the velocity with the components perpendicular to the plane (Fig. 2.10) we can even explain the differences of the CIP and CPP-MR. Similar conclusions have been drawn by Oguchi [29] and Butler *et al.* [30].

The same result was obtained by calculating GMR in the ballistic limit of transport by using the conductance (Eq. (2.36)) [14, 15]. In terms of conduc-

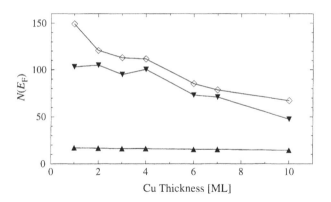

Figure 2.9: Densities of states at the Fermi energy for Co/Cu (100) multilayers as a function of Cu layer thickness [28]; for majority electrons (triangle upwards) and minority electrons (triangle downwards) in P configuration and in AP configuration (diamonds).

tance the GMR ratio is defined as

$$R = \frac{G^{\uparrow} + G^{\downarrow}}{G^{AP}} - 1. \qquad (2.101)$$

The results for $Co_n Cu_m$-multilayers [14] are shown in Fig. 2.11. For the reason given above, i.e., differences in the Fermi velocities, positive GMR ratios are obtained, and the CIP-MR is smaller than CPP-MR.

As discussed in the beginning of this section Fermi velocities are a consequence of the coherent Bragg scattering in the multilayer and are related to

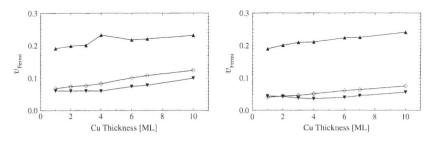

Figure 2.10: The average of the Fermi velocities in the plane of the layers $\sqrt{v_{\parallel}^2}$ (left) and perpendicular to the layers $\sqrt{v_{\perp}^2}$ (right) [28]. For majority (triangle upwards) and minority (triangle downwards) electrons in P and in AP configuration (diamonds).

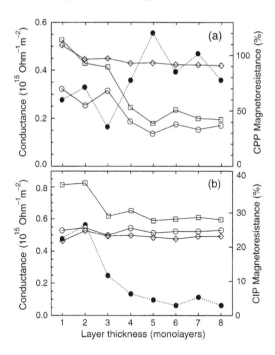

Figure 2.11: The conductance for the majority (diamonds) and the minority (squares) electrons in P configuration and in AP configuration (open circles), and the MR (filled circles) as a function of layer thickness dependence $n$ for (100) oriented $Co_nCu_n$ multilayers; (a) for CPP, (b) for CIP [14].

the transmission and reflection coefficients at the interfaces. Each ML in a unit cell causes an extra Bragg plane in reciprocal space which leads to the opening of superzone gaps in the electronic structure of the multilayer.

The Fermi surface of a free electron gas splits up into nearly cylindrical sheets; these are the so-called minibands (see Fig. 2.12). The number of minibands is equal to the number of MLs. The Fermi surface of a $Co_nCu_m$-multilayer is more complicated but reflects these features (see Fig. 2.13). The majority Fermi surface of the supercell is very similar to the Fermi surface of Cu since the Fermi surfaces of both constituents, Co and Cu, are very similar and consist of only one sheet; the minibands are nearly free electron-like in the center of the zone and cylindrical at the zone boundaries. However, the minority Fermi surface comes from the $d$ bands of Co and the $s$ band of Cu; as a consequence a bunch of minibands is found which contain the peculiarities of the bulk Fermi surfaces. The Fermi surface for the moments of the Co layers oriented antiparallel is more complicated; it contains twice as many minibands because the unit cell is twice as large and the spin channels are mixed.

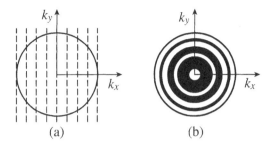

(a)            (b)

Figure 2.12: (a) Cross-section of the Fermi sphere at $k_x = 0$. Dashed lines represent the spacing between the Bragg planes $2\pi/c$ where $c$ is the size of unit cell for the superlattice along its direction of growth, i.e, the $z$ direction. (b) Projection (black area) of the superzone Fermi surface onto the $k_x - k_y$ plane. White rings are superzone gaps that open at the intersection of the Fermi sphere with the Bragg planes.

The supercell Fermi surface of the majority electrons of a $Co_nCu_m$-multilayer in an extended zone scheme (see Fig. 2.14) reflects some general properties of tetragonal systems which influence transport in and perpendicular to the plane of the layers. The Fermi surface in the first Brillouin zone has a cylindrical shape. The flatness of the cylinder is related to the localization of the corresponding electronic states in real space, that is, in one of the layers. This aspect is discussed in the next section in detail. The velocities of states on these sheets (see Eq. (2.12)) are mainly oriented in the plane of the layers and they contribute strongly to the CIP conductivity. The outer sheets of the Fermi surface are shaped like caps and reflect free electron-like behavior

Figure 2.13: Projection of the Fermi surface of $Co_3Cu_7$ (100) multilayers in the plane $k_z = 0$; for majority (left) and minority (middle) electrons in the P configuration, and (right) for the AP configuration.

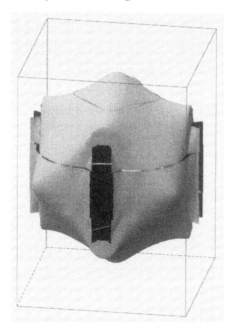

Figure 2.14: The Fermi surface for majority electrons of a $Co_3Cu_7$ (100) multilayer in an extended zone scheme.

of the Bloch functions. Obviously, the velocity vectors point mainly perpendicular to the planes and promote CPP transport. From the topology of the Fermi surface we recognize that transport in and perpendicular to the planes is not related to each other since the conductivity is caused by electrons in different states. Stated differently, CIP and CPP transport properties can be understood as if measurements were made on different samples since different electrons are carrying the current in these different geometries. Similar discussions of the role of the superlattice Fermi surface in CIP and CPP transport have also been made by Oguri *et al.* [31].

### 2.5.2.2 Channeling and quantum confinement

The motion of electrons in potential landscapes characteristic of magnetic multilayers (see Fig. 2.7) leads to quantum confinement [28]. Depending on the height of the potential wells electrons are reflected or transmitted at the interfaces and form states which show a modulation due to the individual layer thicknesses. Some of the states are still free electron-like but most of the states are quantum well-like either in the magnetic or non-magnetic layers.

The eigenstates of a translationally invariant multilayer are all Bloch states $\varphi_0(k, \mathbf{r}, E)$ with a normalized probability amplitude over the unit cell

$$\int d\mathbf{r} |\varphi_0(k, \mathbf{r}, E)|^2 = 1. \tag{2.102}$$

The superposition of all probability amplitudes at a given energy $E$ gives the LDOS or a layerwise decomposed density of states

$$n^\sigma(\mathbf{r}, E) = \sum_{\nu, \mathbf{k}} |\varphi_0(k, \mathbf{r}, E)|^2 \delta(E_k - E) \tag{2.103}$$

which is an important ingredient for understanding conductivity. Since the conductivity is determined by electrons at the Fermi level $E_F$ our interest is focused on LDOS and eigenstates at $E_F$. The LDOS at $E_F$ of the P configuration is shown in Fig. 2.15(a). The local density of states in the majority channel is nearly the same for all MLs. In contrast, the minority electrons are characterized by a very inhomogeneous profile. Due to the $d$ states of Co, the LDOS is much higher in Co than in Cu layers (see also Fig. 2.6). The largest values occur at the Co interfacial layers. This is a general behavior independent of Co or Cu layer thicknesses that can be explained in terms of eigenfunctions and their localization in the superlattice. Electron confinement can be described by means of a layerwise projection of the probability amplitude within the supercell. The analysis leads to four representative states (Fig. 2.15(b-e)). We obtain: extended or free electron-like states with nearly the same probability amplitude in all layers (Fig. 2.15(b)); the so-called quantum well states; states with a pronounced electron confinement in the Co (Fig. 2.15(c)) or Cu layers (Fig. 2.15(d)); and, most surprisingly, states characterized by a high probability amplitude at the Co interface layer as shown in (Fig. 2.15(e)). Some of these states are real interface states with an exponential decay into the Co and Cu layers; but there are also quasi localized states, i.e. resonances, with a high probability amplitude at the Co interface and finite but small probability amplitudes in the middle of the layers.

The spectral weights for these four types of eigenstates are indicated by the corresponding grey scale in Fig. 2.15(a). Due to the uniform potential profile most of the eigenstates in the majority band are extended or free electron-like. In contrast, nearly all minority electrons, while extended in the plane of the layers, are confined perpendicular to the layers. The LDOS of the minority band is dominated by Co quantum well states.

In the AP configuration both spin channels are dominated by quantum well states; all free electron-like states are suppressed. On average, these states are less extended than states in the majority band and less localized than states in the minority band. This immediately implies lower CPP conductance in the AP configuration.

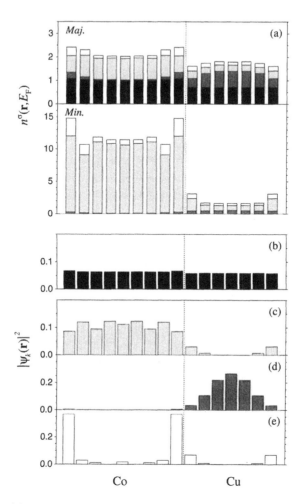

Figure 2.15: (a) Local densities of states at $E_F$ in states/spin Ry. $Co_9Cu_7$ (100) multilayer in P configuration. The different shaded areas correspond to the weights of the four types of eigenstates. Probability amplitudes for suitably chosen $k$ values of: (b) representative extended (majority), (c) Co quantum well (minority), (d) Cu quantum well (minority), and (e) interface (minority) states [28].

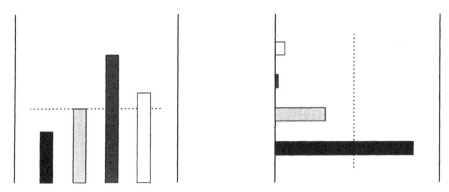

Figure 2.16: Averaged Fermi velocities in a.u. $\sqrt{v_\parallel^2}$ (left) and $\sqrt{v_\perp^2}$ (right) for the classes of eigenstates (corresponding grey scale) discussed in Fig. 2.15 [28]. Dotted line corresponds to the averaged absolute value of the velocity $\sqrt{v^2}$.

The averaged Fermi velocities (Fig. 2.16) of these types of eigenstates nicely elucidate their role in transport. Confined states have their velocity mainly in the plane of the layers; free electron-like states have velocity components in and perpendicular to the plane of the layers. We conclude that CIP transport is mainly driven by quantum well states with a small contribution from extended states. However, CPP transport is mainly due to a few highly conducting free electron-like states. Obviously, by switching the magnetic configuration from AP to P in an external magnetic field the CPP conductivity increases dramatically since highly conducting channels are opened; on the other hand, the CIP conductivity increases only slightly since only a few free electron-like states are added. Consequently the role of the electronic structure of Co/Cu multilayers is to make the CPP larger than the CIP-MR. The same discussion holds for Fe/Cr multilayers.

The phenomenon of quantum confinement in magnetic multilayers can also be viewed as a channeling or waveguide effect [30]. Starting from an electronic structure calculation in a superlattice we can define a two-point or nonlocal conductivity (see Eq. (2.37)). In multilayered systems which consist of stacks of atomic planes with a common underlying 2D periodicity, it is convenient to define a nonlocal layer conductivity, $\sigma_{\mu,\nu}(I, J)$, which is the current in direction $\mu$ in atomic plane $I$ due to an applied electric field of unit magnitude in direction $\nu$ applied in atomic plane $J$ [30]. This is synonymous with the layerwise conductivity defined in Eq. (2.56) except that now one is using a layer index instead of the continuous variable $z$. An example of such a conductivity is shown in Fig. 2.17. We note that the diagonal elements $\sigma_{\mu,\nu}(I, I)$ are the most important contributions; this can readily be understood inasmuch as the two

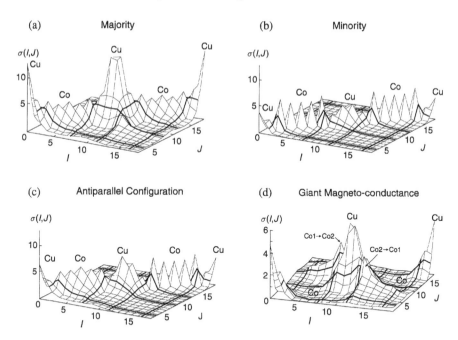

Figure 2.17: Nonlocal layer dependent conductivities, see Eq. (2.104) assuming different scattering rates for Co and Cu layers. (a) Majority spin conductivities for P alignment, (b) minority spin conductivity for P alignment, (c) conductivity for AP alignment, and (d) layer dependent contributions to giant magnetoconductance. Conductivity is in units of $1113/\Omega$cm [30].

point conductivity decays exponentially as a function of the distance between the layers and that the length scale of the decay is controlled by the mean free path. Furthermore, if we would sum up one of the layer indices of the nonlocal conductivity

$$\sigma_{\mu,\nu}(I) = \sum_J \sigma_{\mu,\nu}(I, J) \qquad (2.104)$$

we would end up with a profile of layerwise contributions to CIP conductivity which can be compared to the local densities of states in Fig. 2.15(a); this is tantamount to integrating only over the variable $z$ in Eq. (2.58), and noting the total conductivity is an integral of these $\sigma(z')$ over all $z'$. We note the characteristic concave profile in the Co layers with the largest contributions to the conductivity at the Co interface layers; these are due to the interface states discussed above. On the contrary the profile for the Cu layers is convex with the largest contributions to the conductivity coming from the middle of the layer; this is due to the quantum well states in Cu [28]. The relative contributions to the conductivity from the Cu and Co layers cannot be compared directly, because the above discussion (Fig. 2.15) assumed a constant relaxation time for the entire sample whereas the results of the calculations shown in Fig. 2.17 included different isotropic relaxation times for the Co and Cu layers which were adjusted so as to reproduce the resistivities of thin films of Co and Cu. As a result the Cu layer is highly conducting even though we have a smaller LDOS that contributes to conductivity. The conclusion one can draw is that channeling states play an important role in the conductance of layered structures; this agrees with the result discussed above on the role of quantum confinement. An analogous investigation can be applied to spinvalve structures [34].

While these channeling or quantum confinement effects exist in perfect layered structures they have by and large not been observed in transport phenomena on real multilayers. The reason is that transport averages over all directions in momentum space; for these effects to show up demands that the scattering from the two interfaces bounding a layer are coherent; the roughness of the interfaces between metallic layers and the variations in thickness of the layers do not allow the scattering to add up coherently so that transport does not observe these quantum effects that have been observed in photoemission spectroscopy and interlayer exchange coupling [48, 49].

## 2.5.3    Spin-dependent scattering

### 2.5.3.1    Specular vs diffuse scattering

The sources of scattering in multilayered structures are grouped into those originating within the layers (bulk) and those occurring at the interfaces between layers. The former include atoms of one element that stray into the layer of another element, grain boundaries, and other structural defects that occur in compositionally layered materials. Scattering at interfaces comes from long wavelength geometrical roughness, i.e., compositional disorder of the length scale of the mean free path of the electrons, and from atomic scale roughness due to diffusion of atoms across the interface. These scatter electrons diffusely, inasmuch as there is little correlation between incoming and outgoing momenta after one does impurity averages. It is the primary source of resistance at low temperatures. When transport in periodic multilayered structures, i.e., superlattices, is calculated in the piecewise approach described in Sec. 2.3.3.1, electrons are scattered at perfectly flat interfaces, as described by reflection and transmission coefficients. However, this is specular scattering as electrons retain a memory of their momenta, and is not a source of resistance; however, the difference in the lattice potential in going from one layer to another does affect the scattering coming from impurities. It has been shown that the same impurity scatters differently in the bulk of a layer and at an interface [28].

As first proposed by Fert [36], spin-dependent scattering of conduction electrons, i.e., a different scattering cross-section for spin-up and spin-down electrons, is the primary source of the GMR observed in layered structures. This mechanism is not an intrinsic property of the host metal, but depends on the specific impurity producing the scattering in the corresponding host; in particular, the host potentials where the impurities are embedded differ in the layers or at the interfaces. This scattering is not a source of GMR [11] unless the scattering rate of the electrons is changed by an external field. The relation of the change in scattering rate to the spin-dependent scattering is arrived at through the following line of reasoning. As we discussed above, the scattering of electrons in the spin-up or majority (parallel to the magnetization) con- duction band of transition-metal ferromagnetic alloys, $V_{scatt}^{\uparrow}$, is different from that for electrons in the spin-down or minority band $V_{scatt}^{\downarrow}$. Instead of writing $V_{scatt}^{\uparrow}$ and $V_{scatt}^{\downarrow}$ for the majority and minority electron scattering potentials, one could write it in a rotationally invariant form in spin-$\frac{1}{2}$space, as shown in Eq. (2.9). By choosing the spin axis of quantization for the electron par- allel to the magnetization, one obtains a diagonal $2 \times 2$ matrix for the Pauli spin operator $\sigma$, whose components yield scattering potentials $v \pm j$; these are synonymous with the $V_{scatt}^{\uparrow}$ and $V_{scatt}^{\downarrow}$ introduced earlier. When Eq. (2.9) is ap- plied to bulk ferromagnetic alloys, the application of an external field does not alter the magnetization of a saturated ferromagnet (we are confining ourselves to low temperatures), and the scattering rates are unchanged. In magnetic multilayers and granular films, an external field spatially rearranges the local magnetization $\hat{\mathbf{M}}(\mathbf{r})$ in these structures, and we find from Eq. (2.9) that the scattering potential of the electrons is altered by the field; this change provides a plausible source of GMR.

In making this argument we have assumed that the spin axis of quantization of the electron does *not* follow the local internal field (magnetization). This assumption is justified as long as the time spent by the electron in any one region of magnetization is small compared with the time it takes the electron's axis of quantization to reorient itself parallel to that magnetization. This approximation is valid for layered and granular structures where the size of the regions over which the magnetization changes is small. However, it would be incorrect in cases where the magnetization changes gradually, such as in domain walls [41].

The spin-dependent scattering represented by Eq. (2.9) should be distin- guished from spin-flip scattering off paramagnetic impurities; whereas it is represented by the Hamiltonian-like Eq. (2.9), the classical unit vector repre- senting the magnetization $\hat{\mathbf{M}}$ is replaced by a quantum mechanical spin opera- tor $\mathbf{S}$ representing internal degrees of freedom of a paramagnetic impurity. The distinction between the two is best seen in systems where the magnetization is collinear (either parallel or antiparallel); in this case, one chooses the spin axis of quantization along the magnetization. For spin-dependent scattering there is no spin mixing, i.e., no transitions from spin-up to spin-down or vice versa, because the components of the spin-flip operators $\sigma_x$ and $\sigma_y$, i.e., the magnetization $\hat{M}_x$ and $\hat{M}_y$, are zero for the natural axes. However, for paramagnetic impurities the operators $S_x$ and $S_y$ produce spin-flips. While spin-dependent scattering produces GMR, spin-flip scattering diminishes this effect.

While scattering takes place in both the bulk (layers) and at the interfaces between layers, there is invariably more scattering at the interfaces of metallic structures. Whether this produces spin-dependent scattering depends on the interface magnetism; where it exists one can anticipate strong spin-dependent scattering from the interfacial regions of these structures. Indeed, some treatments of GMR have focused on this mechanism alone, with reasonably good results [50].

In summary, GMR in inhomogeneous magnetic structures has at least two origins: (1) scattering potentials that are different for the spin-up and spin-down conduction bands, i.e., spin-dependent scattering, and (2) conduction band wave functions that depend on the magnetic configuration of the composite structure and determine the effect of impurity scattering on resistivity. The first mechanism can produce GMR by itself (both in CIP and CPP); the second cannot. However, in the presence of any scattering, the second mechanism can cause the spin-dependence of scattering or in case of spin-dependent scattering can amplify or attenuate the MR due to the first mechanism.

### 2.5.3.2    Elastic vs inelastic scattering

At low temperature scattering of electrons is elastic; it conserves the energy of electrons while changing direction of their momenta, and we focused on spin-dependent diffuse scattering as the origin of GMR. At finite temperatures inelastic scattering of electrons by phonons and magnons occurs. All these processes reduce the inelastic mean free path, which is the length scale over which electrons remain in energy eigenstates. Thereby resistivity increases with temperature. If spin dependent processes remained constant, so as to maintain the *difference* between resistivities for ferro- and antiferromagnetic configurations of a multilayer, the increase in the overall resistivity in and of itself reduces the MR ratio. In addition the spin dependence of the scattering mechanisms vary with temperature.

Phonons can conceivably increase MR if one has strongly spin-dependent densities of states to evaluate this scattering, e.g., for Co-based structures [51]. However, in general, they do not produce spin-dependent scattering, and therefore reduce the MR, e.g., Fe/Cr. Magnons reduce GMR in two ways: (1) the spin-dependent scattering decreases with the magnetization and (2) magnons create spin-flips that mix the two spin currents [11]. On balance

GMR invariably decreases with temperature, however MR ratios for Co/Cu multilayers remains considerably closer to their low temperature values than those for Fe/Cr multilayers, so that phonons may be offsetting some of the magnon contributions.

### 2.5.3.3 Impurity scattering

Let us illustrate the points made in the previous two sections by means of impurity scattering. Impurity scattering is the easiest mechanism and most transparent formalism that produces diffusive scattering. Furthermore, impurity scattering can be treated on an *ab-initio* basis from the very beginning through to the transport properties. The general formalism is already known from dilute bulk materials. By means of impurities the translational invariance of the system is broken and we have diffusive scattering; one's knowledge of momentum is lost. An electron in a Bloch state $\varphi_0(k, \mathbf{r}, E)$ with $k = (\nu, \mathbf{k}, \sigma)$ and energy $E_k$ will be scattered into a perturbed Bloch state $\varphi(k', \mathbf{r}, E)$ with energy $E_{k'}$ for which the corresponding probability amplitude is no longer translationally invariant. The scattering process (Fig. 2.18) can be described by the so-called $t$-matrix

$$T_{kk'} = \frac{1}{V} \int d^3r \, \varphi_0(k, \mathbf{r}) V_{scatt}^\sigma(\mathbf{r}) \varphi(k', \mathbf{r}). \qquad (2.105)$$

The transition probability from $k$ to $k'$ can be expressed by means of Fermi's Golden Rule

$$P_{kk'} = \frac{2\pi}{\hbar} cN \, |T_{kk'}|^2 \delta(E_k - E_{k'}). \qquad (2.106)$$

The $\delta$-function means that we restrict our considerations to elastic scattering. This expression is only valid in the case of dilute alloys since it is proportional

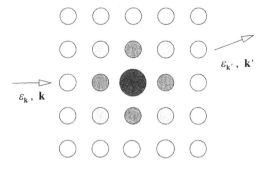

Figure 2.18: Scattering of an incoming Bloch wave $\mathbf{k}$ by a cluster of perturbed potentials induced by an impurity at the center.

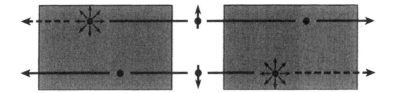

Figure 2.19: Effect of spin-dependent scattering for $\beta \ll 1$ (left) and $\beta \gg 1$ (right).

to the concentration $c$ of impurities in the system. Interactions between defects have not been taken into account (see Eq. (2.105)).

The microscopic transition probability given in Eq. (2.106) is equivalent to the tensor Eq. (2.21). For non-magnetic systems these probabilities are independent of spin. In collinear ferromagnetic systems, the components of Eq. (2.21) depend on the electron's spin. The diagonal elements characterize the spin-dependent scattering cross-section in each spin-channel whereas the non-diagonal elements characterize spin-flip scattering between channels. The off diagonal spin mixing elements result principally from the scattering by spin waves [11] or from collisions between spin-up and spin-down electrons [52], both mechanisms cease to operate at 0 K. But there are also residual spin-mixing terms. Spin-flip can occur when scattering by impurities is influenced by spin-orbit coupling, but the corresponding cross-section is about two orders of magnitude smaller than that of the spin-conserving scattering potential [53].

If the diagonal elements in (spin space) of the microscopic transition probability are summed over all final states the inverse relaxation time (Eq. (2.20)) is obtained. Obviously, the relaxation time depends on state $(\nu, \mathbf{k})$ and spin $\sigma$. To simplify matters, a Fermi surface average of the relaxation time can be considered (Eq. (2.29)). As a result we obtain characteristic relaxation times for each direction of spin $\tau^{\sigma}$, and the spin anisotropy ratio $\beta$ (Eq. (2.31)). If $\beta < 1$ majority electrons of the system are more strongly scattered than the minority (Fig. 2.19).

On the contrary, for $\beta > 1$ minority electrons are more strongly scattered. These differences in relaxation times produce differences in conductivities of the spin channels; one will be highly conducting, the other weakly. This is reflected in the spin anisotropy ratio $\alpha$ (see Eq. (2.27)). Due to the two current model, see Sec. 2.4, and as we now show the total conductivity is determined by the highly conducting channel. As mentioned in Sec. 2.5.2.1, the spin anisotropy in and of itself does not produce GMR; it is the ability to effect spatial changes of the magnetization with an external field that cause GMR. The underlying mechanism is described schematically in Fig. 2.20. Let us consider for example a multilayer with the moments of the magnetic layers

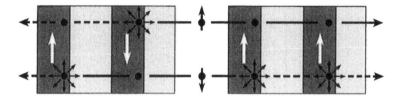

Figure 2.20: Schematic picture of GMR due to spin-dependent scattering. Channels with high and low conductance for P alignment (right) and mixed conductance for both channels in AP alignment (left).

in parallel, and with strong scattering, that is, small relaxation times, in the minority channel. The conductivity of the majority channel is consequently larger than that of the minority channel $\sigma^{P\downarrow} < \sigma^{P\uparrow}$ and the total conductivity $\sigma^P = \sigma^{P\uparrow} + \sigma^{P\downarrow}$ is determined by the faster majority channel. For the antiparallel configuration of the layer magnetizations the scattering strength is switched so that the conductivities $\sigma^{AP\sigma}$ of the two spin channels are equal and satisfy the relation $\sigma^{P\downarrow} < \sigma^{AP\sigma} < \sigma^{P\uparrow}$. Therefore the total conductivity $\sigma^{AP}$ is in general smaller than $\sigma^P$ which would lead to an extrinsic GMR caused by spin-dependent scattering. The measured GMR effect can be understood as the interplay of the intrinsic effect (see Sec. 2.5.1) coming from the electronic structure and this extrinsic effect. As the two are not that closely correlated the intrinsic effect can be amplified, reduced or even inverted by spin-dependent scattering.

### 2.5.3.4 Position-dependent impurity scattering

Before we discuss how one actually determines the scattering rates by using self consistent impurity calculations we point out the influence of the quantum confinement in layered structures on the scattering matrix element. If we assume a spin-dependent $\delta$-function scatterer of finite strength

$$V_{scatt}^{\sigma}(\mathbf{r}_i) = t^{\sigma}\delta(\mathbf{r} - \mathbf{r}_i), \qquad (2.107)$$

the microscopic transition probability, given by Eq. (2.106), will depend on the position of the scatterer $\mathbf{r}_i$ as will the relaxation time

$$(\tau_k(\mathbf{r}_i))^{-1} = 2\pi\,c\,|\varphi_0(k,\mathbf{r}_i)|^2\,n^{\sigma}(\mathbf{r}_i, E_F)\,t^{\sigma 2}. \qquad (2.108)$$

For simplicity, the relaxation time is given in the Born approximation, i.e., we have not considered the effect of the scatterer on the Bloch functions. We see that the relaxation time in Born approximation is determined by the local density of states of the host at the Fermi energy $n^{\sigma}(\mathbf{r}_i, E_F)$ at the impurity

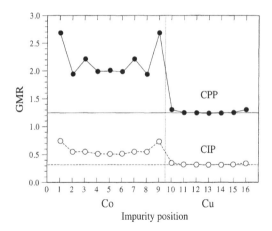

Figure 2.21: GMR of a $Co_9Cu_7$ (100) multilayer produced by a defect with $\beta' = 4$ at different positions in the multilayer: open symbols are for CIP-MR; closed symbols for CPP-MR. The straight lines correspond to the results calculated with isotropic relaxation times, i.e., assuming constant relaxation times throughout the multilayer (see Eq. (2.98)) [28].

site. The corresponding ratio of spin-dependent LDOS's is

$$\gamma(\mathbf{r}_i) = n^{\downarrow}(\mathbf{r}_i, E_F)/n^{\uparrow}(\mathbf{r}_i, E_F) \qquad (2.109)$$

which is always larger one in Co/Cu multilayers. The value at the Co interface is $\gamma = 6.8$. Furthermore, the inverse relaxation time depends on the probability amplitude of the incoming electronic wave function $|\varphi_0(k, \mathbf{r}_i)|^2$ at the impurity site. The spin anisotropy of the scattering strength is simulated by $\beta' = t^{\downarrow}/t^{\uparrow}$. Using Eq. (2.30) for the conductivity it can be shown that the GMR ratio strongly depends on the position of a scatterer in a multilayer. Impurities in the magnetic layers, which can be understood as bulk impurities, increase the GMR ratio (Fig. 2.21); those at the interfaces between layers are especially efficient spin-dependent scatterers because of the interface states that exist there with a high probability amplitude. Impurities in the non-magnetic metallic layer do not produce a difference in the resistivities of the parallel and antiparallel configuration of the multilayer; however they lower the GMR ratio inasmuch as they increase both of these resistivities.

### 2.5.3.5   Self consistent impurity potentials

To obtain the resistivity and GMR caused by $3d$ transition metal impurities in magnetic multilayers several attempts have been made to evaluate the scattering potentials [54]. A complete answer however can only be observed from a

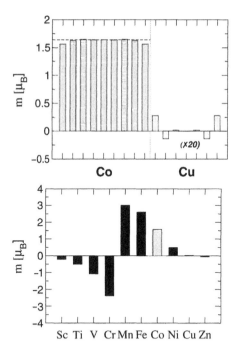

Figure 2.22: Magnetic moments at different sites of a $Co_9Cu_7$ (100) multilayer (left), and for $3d$ transition metal impurities in the same multilayer in the Co interfacial layer (right); the broken line indicates the Co bulk moment [12].

self consistent calculation of the impurity potentials $V_{scatt}^{\sigma}(\mathbf{r})$ in the multilayer. The results presented in this section are performed within a Green's function formalism [5, 12]. As we have indicated above, impurities at interfaces dominate the scattering, therefore we concentrate our attention on these defects. Suppose $3d$ impurity atoms are introduced into a Co/Cu multilayer. The presence of impurities changes the local electronic structure, i.e., the density of states $n^{\sigma}(\mathbf{r}_i, E_F)$ and the corresponding ratio $\gamma$ (Eq. (2.109), Fig. 2.23). As a consequence, specific impurity moments are formed which are either parallel or antiparallel to the Co moments (see Fig. 2.22). The connection between the local electronic structure $\gamma$, the spin anisotropy of scattering $\beta$ and the spin anisotropy of transport $\alpha$ are illustrated in Fig. 2.23. First, all $3d$ transition metal impurities reduce the ratio $\gamma$ (see Eq. (2.109)) at the Co interface site. Second, the spin anisotropy ratio $\beta$ of the relaxation times follows the trend of $\gamma$ but is modulated by the direction of the magnetic moments. Impurities with parallel magnetic moment show an anisotropy $\beta > 1$, whereas impurities with opposite moments show anisotropies $\beta < 1$. The same tendency is reflected in the ratio $\alpha$ for in-plane transport (Fig. 2.24).

For impurities with moments parallel to the magnetization in a layer the conductivity of the majority channel $\sigma^\uparrow$ is larger than that of the minority channel $\sigma^\downarrow$. For impurities whose moments are antiparallel aligned this behavior is reversed. The result is only shown for CIP but is also reflected for CPP configuration [12]. The corresponding GMR results are shown in Fig. 2.25; large positive GMR is obtained for impurities with moments parallel to the layer magnetization, i.e., the effect of the Fermi velocities (intrinsic effect) and of the relaxation time (extrinsic effect) reinforce one another. Systems with opposite impurity moments and spin anisotropy $\beta < 1$ reduce the intrinsic GMR drastically.

The GMR ratios obtained for Fe, Ni and Cu impurities are orders of magnitude larger than observed in Co/Cu superlattices; this is in agreement with calculations for NiFe/Cu superlattices [55, 56]. Undoubtedly, this ratio will be reduced when other scattering mechanisms, e.g., scattering arising in the bulk of the layer, are taken into account. Probably, the key to understanding the discrepancies in these systems is the neglect of spin-flip scattering. Because of this reason conduction in the fast channel is overestimated and nearly produces a short circuit [5, 40].

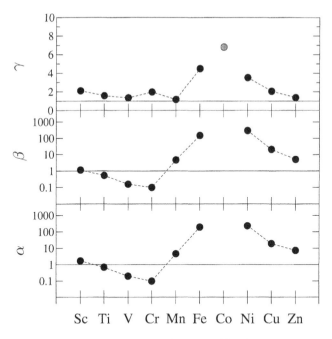

Figure 2.23: Spin anisotropy ratios $\gamma$, $\beta$ and $\alpha$ (CIP) for $3d$ transition metal impurities in a $Co_9Cu_7$ (100) multilayer in the Co interfacial layer [12]; the grey circle corresponds to the host value of $\gamma$.

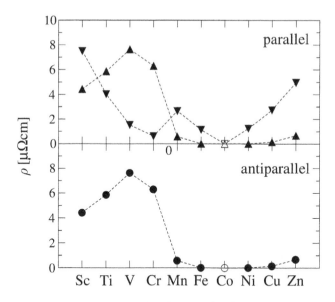

Figure 2.24: Residual resistivities of a $Co_9Cu_7$ (100) multilayer due to 3d transition metal impurities in the Co interfacial layer: for the majority (triangles upwards) and minority bands (triangles downward) in P alignment and AP alignment (circles) of the Co moments in adjacent layers [12] .

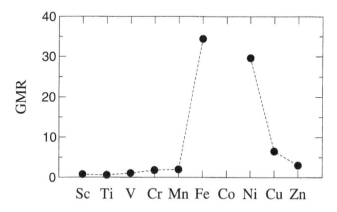

Figure 2.25: CIP-MR caused by 3d transition metal impurities in the Co interfacial layer of a $Co_9Cu_7$ (100) multilayer.

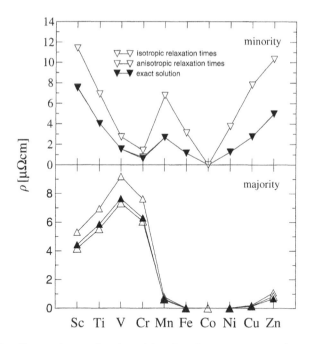

Figure 2.26: Comparison of resistivities for the majority and minority electrons of a $Co_9Cu_7$ (100) multilayer with $3d$ transition metal impurities in the Co interfacial layer; exact solution (Eq. (2.25)) in comparison to several approximations: anisotropic relaxation times (Eq. (2.32)) and isotropic relaxation times (Eq. (2.30)).

The results presented above have been obtained by solving the Boltzmann equation (Eq. (2.19)) and calculating the conductivity by means of Eq. (2.25). Figure 2.26 shows a comparison of the exact solution Eq. (2.25) with the approximations (Eqs. (2.30) and (2.32)). Obviously, the exact solution agrees very well with the approximation (Eq. (2.32)) when the anisotropic relaxation times are included. In contrast, isotropic relaxation times reproduce the right qualitative behavior; however, the absolute values can be off by more than 50%.

### 2.5.3.6    Interdiffusion and alloying

Although the impurity results discussed above are instructive, inasmuch as they elucidate the microscopic processes behind the final result for the conductivity and GMR, the dilute limit is far from the experimental situation when samples have lots of defects. A step in the direction of an *ab-initio* as-

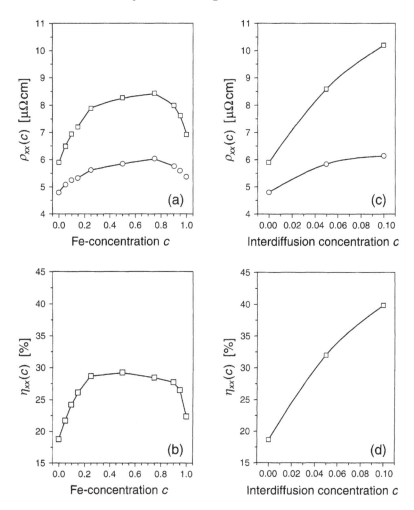

Figure 2.27: CIP resistivities $\rho_{xx}(c)$ and CIP-MR $\eta_{xx}(c)$ for the multilayer system $(Co_3Cu_3)_5$ embedded in fcc Cu(100) for the P (circles) and AP (squares) configurations as a function of the concentration $c$ of impurities. (a) and (b) refer to homogeneous alloying in the Co layers with Fe impurities; (c) and (d) to interdiffusion at the Co/Cu interfaces [39].

sessment for more realistic systems are studies of magnetic multilayers with regions of concentrated substitutional alloys, i.e., interdiffusion at interfaces and alloying of the layers. This was recently carried out for Co/Cu multilayers by means of a spin-polarized relativistic KKR method for layered systems together with the coherent potential approximation to take into account effects

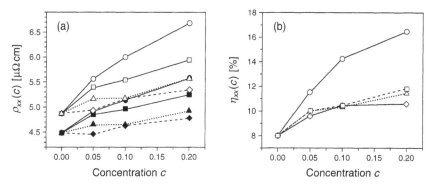

Figure 2.28: Comparison of the effects of interdiffusion and homogeneous alloying on (a) the CIP resistivities $\rho_{xx}(c)$, and (b) CIP-MR $\eta_{xx}(c)$ for the multilayer system $(\mathrm{Co_3Cu_3})_5$ embedded in fcc $\mathrm{Cu(100)}$ as a function of the concentration. Circles refer to interdiffusion at the Co/Cu interfaces; diamonds, triangles and squares to homogeneous alloying in the Co layers with Fe, Ni, and Cu, respectively. Full symbols correspond to the P configuration, open symbols to the AP configuration for moments of adjacent Co layers [39].

of interdiffusion and alloying. The electrical transport properties were calculated within the Kubo-Greenwood formalism neglecting the scattering-in term, i.e., "vertex corrections" [39]. Among other things it was nicely demonstrated that: CIP-MR increases linearly with the number of repetitions of the unit cell, i.e., the $\mathrm{Co}_n\mathrm{Cu}_m$ motif, until it saturates; and the resistivity and GMR can be increased by interdiffusion (see Figs. 2.27(c) and (d)). This latter result agrees with those obtained in the dilute limit, where Cu impurities in the Co layers lead to an enhanced GMR ratio [12]. The role of homogeneous alloying the Co layers with Fe is shown in Figs. 2.27(a) and (b). By starting from pure Co layers the CIP-MR can be increased by alloying; it saturates for concentrated alloys, and then decreases linearly. A comparison of the role of interdiffusion with homogeneous alloying is insightful; due to the presence of interface states interdiffusion produces stronger spin-dependent scattering than homogeneous alloying and it acts much more efficiently to enhance CIP-MR; see Fig. 2.28, Sec. 2.5.3.4, and [28].

## 2.5.4   Finite temperature

In addition to creating inelastic processes and reducing the inelastic mean free path of electrons, finite temperatures broaden the step in the Fermi function and allow electron states away from the Fermi surface to participate in conduction. Density of states (DOS) vary with energy; in particular in magnetic systems the spin dependence of the DOS is variable, and can either increase or decrease with energy. The broader range of electron states that participate in conduction at finite temperatures alters the rates at which they are scattered as well as the spin dependence of the scattering, and concomitantly the GMR. While inelastic processes almost invariably reduce GMR, depending on the electronic structure of the multilayer, finite temperature broadening of the Fermi function can either increase or decrease GMR.

## 2.6   Relation of GMR to TMR

In the next chapter, the MR of tunnel junctions (TMR) with magnetic electrodes is discussed; here we contrast some aspects of TMR to GMR. While we discussed generic multilayers, one category is spin valves: essentially trilayers consisting of two magnetic layers with a non-magnetic metallic spacer. Magnetic tunnel junctions (MTJ's) consist of the same type of magnetic electrodes (metallic layers), but with an insulating or semiconducting spacer. While metallic multilayers are studied in both CIP and CPP, CPP is the only geometry of interest for tunnel junctions. Their resistance is considerably higher, so that for comparable MR ratios, $R$, the change in voltage as one reorients the magnetization of one electrode relative to the other is considerably larger for MTJ's than in spin valve or multilayered structures. On the other hand, the current in MTJ's is tiny compared to that in metallic multilayers, so that the latter would be preferred in devices that are current driven.

The factors that influence TMR and GMR are quite different. Conduction for CPP is controlled by those regions that limit the current; in MTJ's they are the insulating spacer and electrode/insulator interfaces; their resistivities are huge compared to the metallic electrodes. In multilayers no one region stands out as the resistivities of the various metallic layers are comparable, and current is controlled by the electronic structure and some distribution of scatterers throughout the structure. As the resistivities of the electrodes are tiny compared to the insulator, scattering in them do not affect the current. While spin-dependence of electronic structure and scattering in the whole structure are responsible for GMR, these factors as they pertain to the electrodes do not play a role in TMR. In MTJ's the MR is controlled by the spin-dependence of conduction at the electrode/insulator interface, and by the spin-dependence of the band structure in the insulator. To date most of the insulators which have been successfully used in MTJ's, are non-magnetic, e.g., $Al_2O_3$, $SrTiO_3$, so that interfaces control their MR.

The screening length in the metallic electrodes is short, of the order of an angstrom, so that the DOS is quite local and can vary considerably from one region to another; specifically it is different at the surface or interface from that in the bulk of an electrode. A striking example is Fe: in the bulk $\mu_{Fe} \sim 2.2\mu_B$ and the DOS for majority electrons is higher than minority on the Fermi surface (one calls this positive spin polarization); at the surface $\mu_{Fe} \sim 3\mu_B$ and the DOS in negatively spin polarized [57]. In addition, the DOS and its spin polarization depend on the crystallographic orientation of the surface or interface. The bonding of the orbitals on the electrode surface to those in the insulator, and its dependence on spin is also crucial for TMR. So it is the spin dependence of the magnetic electrode's surface DOS, and the bonding to the insulator that control the MR in MTJ's. In some metallic multilayers, the spin-dependent interfacial scattering has been singled out as the primary cause of GMR; this occurs when the magnetic layers are thin so that the contribution from scattering in the bulk does not contribute substantially. Even in such systems the role of interfaces does not play as decisive a role in GMR as in TMR for the following reason. The metal electrode/insulator interface is highly reflective (transmission is low) so that electrons spend an inordinate time there. Therefore, the electronic structure and scattering in the interfacial regions of MTJ's are the factors that control their MR. The interface between two metals is far more transmissive than reflective; even if spin-dependent scattering at the interfaces is large, that in the bulk of the layers play some role in the overall conduction. Also, while we can neglect scattering in electrodes one should not model them as coherent conductors. Doing this will lead to erroneous conclusions about the origins of MR in MTJ's [58].

Electron conduction in metallic multilayers is mostly in the linear response region and therefore largely controlled by states at the Fermi surface. However, for tunnel junctions it is possible to maintain a finite voltage drop across the insulating barrier, and thereby inject electrons at energies $eV$ above the Fermi level. Therefore in MTJ's one probes the energy and spin dependence of the interfacial DOS about the Fermi surface [59]. When hot electrons give up their energy through inelastic scattering by surface magnons at the electrode interface they affect the TMR; spin-flip scattering at the interface is particularly effective in reducing the MR of MTJ's [60]. Finally, while the resistance of metallic multilayers increase with temperature, due to increased scattering, it decreases for tunnel junctions because tunneling conduction is controlled by the number of available states which increases with temperature.

# References

[1] M. N. Baibich, J. M. Broto, A. Fert, F. Nguyen Van Dau, F. Petroff, P. Etienne, G. Creuzet, A. Friederich and J. Chazelas, Phys. Rev. Lett. **61**, 2472 (1988); G. Binash, P. Grünberg, F. Saurenbach and W. Zinn, Phys. Rev. B **39**, 4828 (1989).

[2] R. Schad, C. D. Potter, P. Beliën, G. Verbanck, V. V. Moschalkov and Y. Bruynseraede, Appl. Phys. Lett. **64**, 3500 (1994).

[3] C. Vouille, A. Fert, A. Barthélémy, S. Y. Hsu, R. Loloee and P. A. Schroeder, J. Appl. Phys. **81**, 4573 (1997), and C. Vouille, A. Barthélémy, F. Elokan Mpondo, A. Fert, P. A. Schroeder, S. Y. Hsu, A. Reilly, R. Laloee, Phys. Rev. B **60**, 6710 (1999).

[4] See for example, N. W. Ashcroft and N. D. Mermin, *Solid State Physics* (Holt, Rinehart and Winston, New York, 1976) Chapters 1 & 2.

[5] I. Mertig, Rep. Prog. Phys. **62**, 123 (1999).

[6] R. Landauer, IBM J. Res. Dev. **1**, 223 (1957); Z. Phys. B **21**, 247 (1975); J. Phys. Cond. Matter **1**, 8099 (1989); M. Büttiker, IBM J. Res. Dev. **32**, 317 (1988).

[7] Supriyo Datta, *Electronic Transport in Mesoscopic Systems* (Cambridge University Press, Cambridge, England, 1995), see pp. 151-157.

[8] P. Hohenberg and W. Kohn, Phys. Rev. **136**, B 864 (1964); W. Kohn and L. J. Sham, Phys. Rev. **140**, A 1133 (1965); L. J. Sham and W. Kohn, Phys. Rev. **145**, 561 (1966).

[9] J. M. Ziman, *Electrons and Phonon* (Oxford University Press, London, 1962).

[10] N. F. Mott, Adv. Phys. **13**, 325 (1964).

[11] A. Fert, J. Phys. C **2**, 1784 (1969); A. Fert and I. A. Campbell, J. Phys. F **6**, 849 (1976); I. A. Campbell and A. Fert, in *Ferromagentic Materials* Vol. **3**, edited by E. P. Wohlfarth (North Holland, Amsterdam, 1982), p. 769.

[12] J. Binder, P. Zahn and I. Mertig, Phil. Mag. B **78**, 537 (1998); J. App. Phys. **87**, 5182 (2000); Phys. Rev. B (to be published).

[13] Yu. V. Sharvin, Zh. Eksp. Teor. Fiz. **48**, 984 (1965).

[14] K. M. Schep, P. J. Kelly and G. E. W. Bauer, Phys. Rev. Lett. **74**, 586 (1995); Ph.D. thesis (Technische Universität Delft, Delft, 1997).

[15] M. A. M. Gijs and G. E. W. Bauer, Adv. in Phys. **46**, 285 (1997).

[16] R. Kubo, M. Toda and N. Hashitsume, *Statistical Physics II: NonEquilibrium Statistical Mechanics*, Springer Series in Solid-State Sciences, Vol. **31** (Springer-Verlag, Berlin 1985), Chaps. 4 and 5; R. Kubo, J. Phys. Soc. Japan **12**, 570 (1957); also see Mahan.

[17] G. D. Mahan, *Many-Particle Physics* (Plenum Press, New York, 1981); see Secs. 3.7, 7.1 and 7.3.

[18] G. Rickayzen, *Green's Functions and Condensed Matter* (Academic Press, London, 1980); see in particular Chap. 4.

[19] H. E. Camblong, P. M. Levy and S. Zhang, Phys. Rev. B **51**, 16052 (1995); H. E. Camblong *et al.*, Phys. Rev. B **47**, 4735 (1993).

[20] S. Doniach and E. H. Sondheimer, *Green's Functions for Solid State Physicists* (Benjamin/Cummings, Reading, 1974), Chap. 5.

[21] T. Valet and A. Fert, Phys. Rev. B **48**, 7099 (1993).

[22] D. S. Fisher and P. A. Lee, Phys. Rev. B **23**, 6851 (1981).

[23] R. E. Camley and J. Barnas, Phys, Rev. Lett. **63**, 664 (1989).

[24] R. Q. Hood and L. M. Falicov, Phys. Rev. B **46**, 8287 (1992).

[25] J. Barnás and A. Fert, Phys. Rev. B **49**, 12835 (1994).

[26] S. Zhang and P. M. Levy, Phys. Rev. B **57**, 5336 (1998).

[27] P. M. Levy, Solid State Physics, Vol. **47**, edited by H. Ehrenreich and D. Turnbull (Academic Press, Cambridge, MA, 1994), pp. 367-462.

[28] P. Zahn, I. Mertig, M. Richter and H. Eschrig, Phys. Rev. Lett. **75**, 2996 (1995); P. Zahn, J. Binder, I. Mertig, R. Zeller and P. H. Dederichs, Phys. Rev. Lett. **80**, 4309 (1998); I. Mertig, P. Zahn, M. Richter, H. Eschrig, R. Zeller and P. H. Dederichs, J. Mag. Mag. Mater. **151**, 363 (1995).

[29] T. Oguchi, J. Magn. Magn. Mat. **126**, 519 (1993).

[30] W. H. Butler, J. M. MacLaren and X.-G. Zhang, Mat. Res. Soc. Proc. Vol. **313**, 59 (1993); W. H. Butler, X.-G. Zhang, D. M. C. Nicholson, T. C. Schulthess and J. M. MacLaren, Phys. Rev. Lett. **76**, 3216 (1996).

[31] A. Oguri, Y. Asano and S. Maekawa, J. Phys. Soc. Japan **61**, 2652 (1992); Y. Asano, A. Oguri and S. Maekawa, Phys. Rev. B **48**, 6192 (1993)

[32] L. Szunyogh, B. Újfalussy, P. Weinberger and J. Kollár, Phys. Rev. B **49**, 2721 (1994), J. Phys. Cond. Matter **6**, 3301 (1994); L. Szunyogh, B. Újfalussy and P. Weinberger, Phys. Rev. B **51**, 9552 (1995); B. Újfalussy, L. Szunyogh and P. Weinberger, Phys. Rev. B **51**, 12836 (1995).

[33] R. Zeller, P. H. Dederichs, B. Újfalussy, L. Szunyogh and P. Weinberger, Phys. Rev. B **52**, 8807 (1995); R. Zeller, Phys. Rev. B **55**, 9400 (1997); P. Zahn, I. Mertig, R. Zeller and P. H. Dederichs, Phil. Mag. B **78**, 411 (1998).

[34] W.H. Butler, X.-G. Zhang, D. M. C. Nicholson and J. M. MacLaren, J. Appl. Phys. **76**, 6808 (1994); Phys. Rev. B **52**, 13399 (1995); W. H. Butler, X.-G. Zhang, D. M. C. Nicholson, J. M. MacLaren, V. S. Speriosu and B. A. Gurney, Phys. Rev. B **56**, 14574 (1997).

[35] P. Weinberger, P. M. Levy, J. Banhart, L. Szunyogh and B. Újfalussy, J. Phys. Cond. Matter **8**, 7677 (1996).

[36] M. N. Baibich, J. M. Broto, A. Fert, F. Nguyen Van Dau, F. Petroff, P. Etienne, G. Creuzet, A. Friederich and J. Chazelas, Phys. Rev. Lett. **61**, 2472 (1988).

[37] J. Bass, Q. Yang, S. F. Lee, P. Holody, R. Loloee, P. A. Schroeder and W. P. Pratt Jr., J. Appl. Phys. **75**, 6699 (1994); Q. Yang, S. F. Lee, P. Holody, R. Loloee, P. A. Schroeder, W.P. Pratt Jr. and J. Bass, Phys. Rev. Lett. **72**, 3274 (1994).

[38] L. Piraux, S. Dubois and A. Fert, J. Mag. Mag. Mater. **159**, L287 (1996); L. Piraux, S. Dubois, J. L. Duvail, K. Ounadjela and A. Fert, J. Mag. Mag. Mater. **175**, 127 (1997); L. Piraux, S. Dubois, A. Fert and L. Belliard, Europhys. J. B **4**, 413 (1998).

[39] C. Blaas, P. Weinberger, L. Szunyogh, P. M. Levy and C. B. Sommers, Phys. Rev. B **60**, 492 (1999).

[40] W. H. Butler, X.-G. Zhang, T. C. Schulthess, D. M. C. Nicholson, A. B. Oparin and J. M. MacLaren, J. Appl. Phys. **85**, 5834 (1999).

[41] P. M. Levy and S. Zhang, Phys. Rev. Lett. **79**, 5110 (1997).

[42] U. Rüdiger, J. Yu, S. Zhang, A. D. Kent and S. S. P. Parkin, Phys. Rev. Lett. **80**, 5639 (1998); U. Rüdiger, J. Yu, L. Thomas, S. S. P. Parkin and A. D. Kent, Phys. Rev. B **59**, 11914 (1999).

[43] S. Zhang, P. M. Levy and A. Fert Phys. Rev. B **45**, 8689 (1992).

[44] See Ref. 27, pp. 437-440.

[45] P. Dauguet, P. Grandit, J. Chaussy, S. F. Lee, A. Fert and P. Holody, Phys. Rev. B **54**, 1083 (1996); A. Vedyayev, N. Ryzhanova, B. Dieny, P. Dauguet, P. Grandit and J. Chaussy, Phys. Rev. B **55**, 3728 (1997).

[46] A. Barthélémy, A. Fert and F. Petroff, *Handbook of Magnetic Materials* Vol. **12**, edited by K. H. J. Buschow (Elsevier Science, Amsterdam, The Netherlands, 1999) Chap. 1; see in particular Table 1; J. Bass, W. P. Pratt and P. A. Schroeder, Comments Cond. Mater. Phys. **18**, 223 (1998), see Table 1.

[47] S. D. Steenwyk, S. Y. Hsu, R. Loloee, J. Bass and W. P. Pratt, Jr., J. Appl. Phys. **81**, 4011 (1997), J. Mag. Mag, Mater. **170**, L1 (1997).

[48] J.E. Ortega and F. J. Himpsel, Phys. Rev. Lett. **69**, 844 (1992).

[49] P. Lang, L. Nordström, R. Zeller and P. H. Dederichs, Phys. Rev. Lett. **71**, 1927 (1993).

[50] S. S. P. Parkin, A. Modak and D. J. Smith, Phys. Rev. B **47**, 9136 (1993).

[51] B. Loegel and F. Gautier, J. Phys Chem. Solids **32**, 2723 (1971).

[52] A. Bourquart, E. Daniel and A. Fert, Phys. Lett. **26 A**, 260 (1968).

[53] P. Monod, Ph.D. thesis (Universite Paris-Sud, Paris, 1968).

[54] J. Inoue, A. Oguri and S. Maekawa, J. Phys. Soc. Japan **60**, 376 (1991); H. Itoh, J. Inoue and S. Maekawa, Phys. Rev. B **47**, 5809 (1993).

[55] W. H. Butler, J. M. MacLaren and X.-G. Zhang, Mat. Res. Soc. Sym. Proc. **313**, 59 (1993).

[56] J. M. MacLaren, S. Crampin, D. D. Vvedensky and J. B. Pendry, Phys. Rev. B **40**, 12164 (1989).

[57] K. Wang, S. Zhang, P. M. Levy, L. Szunyogh and P. Weinberger, J. Mag. Mag. Mater. **189**, L131 (1998); K. Wang, PhD. Thesis, New York University (1998).

[58] S. Zhang and P. M. Levy, Phys. Rev. Lett. **81**, 5660 (1998).

[59] J. M. De Teresa, A. Barthélémy, A. Fert, J. P. Contour, R. Lyonnet, F. Montaigne, P. Seneor and A. Vaurès, Phys. Rev. Lett. **82**, 4288 (1999).

[60] S. Zhang, P. M. Levy, A. C. Marley and S. S. P. Parkin, Phys. Rev. Lett. **79**, 3744 (1997).

# Chapter 3

# Experiments of Tunnel Magnetoresistance

*Terunobu Miyazaki*

## 3.1 Introduction

During the past 20 years much effort has been made for better understanding of the physics of magnetism in ultrathin films and multilayers. One great reason for this is due to continuous progress in thin film fabrication techniques such as electron beam evaporation, sputtering and molecular beam epitaxy (MBE). In addition to the theoretical reports on the enhancement of magnetic moment at the surfaces of ultrathin films and/or interesting and noble magnetic properties arising from interfaces and/or nanostructures have attracted the attention of researchers. Table 3.1 summarizes the new findings in the field of magnetic thin films which have influenced much the development of their applications. Until 1988, the main subjects of the research on magnetic properties were relatively restricted to the perpendicular magnetic anisotoropy and magneto-optical properties. Discovery of the giant magnetoresistance (GMR) effect in Fe/Cr supperlattice in 1988 completely changed the focus of magnetic thin film research to magnetotransport phenomena. Giant tunnel magnetoresistance effect at room temperature (to be explained in detail in this chapter) was reported 7 years after the discovery of the GMR effect. Developments in experimental studies on ultrathin films and interface before 1991 can be found in the review reports summarized by Falicov *et al.* (1990) [1], Bader (1990) [2], Gradmann (1991) [3], Shinjo (1991) [4], and Siegmann (1992) [5]. Reports on the development of the GMR effect before 1994 have been summarized by Miyazaki [6] and more recent reports cited in Chapter 3 of this book. Reports on magnetic nanostructures research before about 1998 were nicely organized by Himpsel *et al.* [7] in their review article. This chapter is presented as follows: Section 3.1.2 reviews the beginning of the study of tunnel magnetoresistance (TMR) effect and also various kinds of magnetic tunnel junctions. Temperature dependence of TMR ratio and bias voltage dependence of TMR are explained in Sec. 3.1.3. Section 3.1.4 describes the recent progress of TMR study related to the application of TMR effect to magnetic read heads and magnetic random access memories (MRAM).

Table 3.1: Topics or new findings in the field of magnetic thin films and/or magnetic nanostructure past about 20 years

| | |
|---|---|
| 1980 | • Enhancement of surface magnetic moment for ultrathin films |
| ≀ | • Perpendicular magnetic anisotropy for artificial supperlattices or multilayers |
| | • Enhancement of Kerr rotation angle for superlattices |
| | • Development of extremely soft magnetic thin films |
| 1988 | • Finding a giant magnetoresistance in Fe/Cr superlattices |
| 1989 | • Finding an oscillation of interlayer exchange coupling |
| | • Finding a soft GMR materials at room temperature |
| 1995 | • Finding a giant tunnel magnetoresistance effect at room temperature |

# 3.2   Fabrication process and various kinds of magnetic tunnel junctions

The pioneering work on TMR was done by Julliere [8] in 1975. He formed a Fe/Ge-oxide/Co junction and measured the conductance at $T \leq 4.2\,\mathrm{K}$ in a magnetic field as a function of applied voltage. At zero bias, the relative change of conductance, $\Delta G/\bar{G}$, was approximately 14%, where $\Delta G$ is the difference between the two values corresponding to parallel and antiparallel magnetization configurations of the two ferromagnetic films respectively and $\bar{G}$ is the average of the two conductance. This change in conductance decreased rapidly with increasing voltage and became about 2% at 6 mV. Seven years later, Maekawa and Gäfvert [9] reported the MR ratio for Ni/NiO/ferromagnet with Ni, Co, or Fe as the ferromagnetic electrode. The TMR ratio was below several percent even low at temperatures below 4.5 K. After that several results [10, 11] were reported for ferromagnet/insulator/ferromagnet junctions. The TMR ratio at room temperature was around 1%. Miyazaki *et al.* [12] demonstrated a nice correlation between the MR and magnetization curve in an 82NiFe/Al-oxide/Co tunnel junction and obtained a relatively large TMR ratio of even at room temperature. Yaoi *et al.* [13] reported the dependence of the tunnel conductance on the relative angle $\theta$ between the magnetization vector of the two magnetic layers. The conductance was well expressed in the form $G = G_0(1 + \cos\theta)$. They also studied the dependence of the conductance on temperature and applied voltage and found a zero bias anomaly at low temperatures.

The finding of large TMR ratio at room temperature [14,15] accelerated the study of spin-dependent transport in magnetic nanostructures. Recently there has been a great interest for the application of the TMR effect on magnetic read head in hard disk drive and magnetoresistive random access memory which are described in detail in Chapter 5.

**(a) Fabrication of magnetic tunnel junctions**   In the early stage of TMR study there has been a problem of making a barrier that was uniform, free from pinholes and thin enough to allow electron tunneling. The magnetic layers were prepared using electron beam evaporation or sputter deposition. Aluminum oxide has been the most useful and reliable material for preparing the insulator barrier because it ($Al_2O_3$ or $AlO_x$) is chemically stable besides being relatively uniform and possibly pinhole-free. In this case, pure aluminum was also deposited using electron beam evaporation or sputtering. The aluminum oxide layer was fabricated by natural oxidization or plasma oxidization. The thickness, uniformity and content of oxygen (the atomic ratio between oxygen and aluminum) of the aluminum oxide layer influence much on the TMR properties of the tunnel junctions. The most popular method used to evaluate the tunnel barrier properties is to fit the current ($I$) vs voltage ($V$) curve into Simmons' relation in order to obtain the effective barrier height and width of the junctions (see Sec. 3.1.4). Similar to the case of the progress in studying the GMR effect in artificial superlattices, magnetic and transport properties for the magnetic tunnel junctions have been investigated for various kinds of junction structures. Three type of junctions: trilayer standard ferromagnet (F)/insulator (I)/ferromagnet (F) junctions, spin-valve-type junctions with an exchange-biased layer, and double or triple barrier layers junctions have been reported. First, we summarize the characteristics of the typical junctions mentioned and the main results for each junction. After that, We will briefly mention the TMR effect observed in granular materials.

**(b) Ferromagnet/insulator/ferromagnet trilayer junctions**   This type of junction is a standard one and most of the TMR studies have been carried out on this type since the first report of the TMR effect published by Julliere [8]. For the ferromagnets of both electrodes, pure Fe [8, 14, 16] Co [13, 17, 18], Ni [9, 17] metals, $Fe_{1-x}Ni_x$[19], Fe-Co [15] and 80Ni-Fe [12, 13, 18] were used. A popular combination of the electrodes is 80Ni-Fe (permalloy, Py) and pure Co, because permalloy exhibits a soft magnetic property and Co a semi-hard magnetic property and we are able to obtain an antiparallel state of magnetizations due to their different coercivities. For trilayer junction, various kinds of insulating barrier such as NiO, $Al_2O_3$, AlN, MgO and $HfO_2$ have been investigated and Al-oxide is the most popular one as described above.

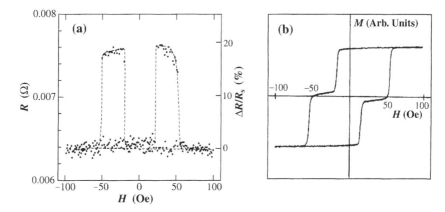

Figure 3.1: (a) Magnetoresistance via magnetic tunneling and (b) magnetic hysteresis curve in an Fe/Al$_2$O$_3$/Fe junction at room temperature. The resistance is low when the two Fe layers are magnetized parallel (for $|H| < 20$ Oe and $|H| > 52$ Oe) and high in between, as seen in conjunction with the $M(H)$ hysteresis curve. The different switching field of the two Fe layers were generated by deposition at room temperature (high coercivity) and 200 °C (low coercivity).

Figure 3.1 shows a typical example of the resistance as a function of magnetic field at room temperature for a Fe/Al$_2$O$_3$/Fe trilayer junction and the corresponding magnetic hysteresis curve. The resistance increased sharply at ±20 Oe and decreased slightly at about ±50 Oe followed by a rapid decrease with further increase of the magnetic field corresponds well with that of the magnetic hysteresis curve shown in Fig. 3.1(b). From the practical point of view, the magnetoresistance ratio is conventionally defined as $\Delta R/R_s = (R_{AP} - R_P)/R_P$, where $R_s$ is the resistance at saturation of magnetization and $R_P$ ($R_{AP}$) is the resistance for the parallel (antiparallel) direction of magnetization in the two electrodes. Hereafter, we call this value as TMR ratio. Some of the trilayer junctions reported before 1999 are summarized in Table 3.2. The last junction described is very interesting, because its Fe layer is a single crystal with three different crystal orientations. The exact stacking is MgO(001)/Cr(001) (400 Å)/Au(001) (1000 Å)/Fe(001) (100 Å)/Al$_2$O$_3$ (10 ∼ 30 Å)/Fe$_{50}$Co$_{50}$/Au(5000 Å) for (001) oriented Fe single crystal case. In order to vary the orientation, the Cr layer was changed to another materials. The Fe electrodes with (100), (110) and (211) planes were investigated. One more point that should be noted in this experiment is that the amorphous Al$_2$O$_3$ tunnel barrier was directly prepared on the Fe single crystal electrode by a reactive deposition method. Figure 3.2 is the TMR ratio at 293 K (a) and

Table 3.2: Ferromagnet/insulator/ferromagnet trilayer junctions reported

| Tunnel junction | Year | Authors |
|---|---|---|
| Fe/GeO/Co | 1975 | Julliere [8] |
| Ni/NiO/M | 1982 | Maekawa and Gäfvert [9] |
| (M=Ni, Co, Fe) | | and Suezawa and Gondo [10] |
| Fe-C/$Al_2O_3$/Fe-Ru | 1990 | Nakatani and Kitada [11] |
| 80NiFe/$Al_2O_3$/Co | 1991 | Miyazaki *et al.* [12, 13] |
| M/$GdO_x$/Fe | 1992 | Nowak and Rauluszkiewicz [16] |
| (M=Gd, Fe) | | |
| 81NiFe/$Al_2O_3$/Co | 1994 | Plaskett *et al.* [18] |
| Fe/$Al_2O_3$/Fe | 1995 | Miyazaki and Tezuka [14] |
| CoFe/$Al_2O_3$/Co | 1995 | Moodera *et al.* [15] |
| $Fe_{0.7}Pt_{0.3}$/$Al_2O_3$/NiFe | 1996 | Moodera and Kinder [20] |
| Fe/$Al_2O_3$/$Fe_xNi_{1-x}$ | 1996 | Tezuka and Miyazaki [19] |
| M/O/M | 1997 | Platt *et al.* [21] |
| (M=Co, Fe, CoFe, | | |
| O=CoO, NiO, $Ta_2O_5$, | | |
| MgO, $HfO_2$) | | |
| NiFeCo/$Al_2O_3$/CoFe | 1997 | Beech *et al.* [22] |
| NiMnSb/$Al_2O_3$/$Ni_{0.8}Fe_{0.2}$ | 1997 | Tanaka *et al.* [23] |
| NiMnSb/$Al_2O_3$/$Ni_{0.5}Fe_{0.5}$ | | |
| Co/$Al_2O_3$/M | 1997 | Moodera *et al.* [24] |
| (M=$Co_{50}Fe_{50}$, $Ni_{80}Fe_{20}$) | | |
| Fe/$AlO_x$/$Co_{50}Fe_{50}$ | 1997 | Tsuge and Mitsuzuka [25] |
| Co/$Al_2O_3$/Co | 1998 | Oepts *et al.* [26] |
| M1/Al-$AlO_x$/M2 | 1998 | Zang and White [27] |
| (M1=NiFe, NiFeCo, | | |
| M2=Co, CoFe) | | |
| Fe(single)/$Al_2O_3$/$Fe_{50}Co_{50}$ | 1999 | Yuasa*etal.* [28] |

2 K (b) as a function of barrier thickness. As seen in the figure the TMR ratio strongly depends on the crystal orientation of the Fe electrode. The average value of the TMR ratio at 2 K is 13% for Fe (001), 32% for (110) and 42% for Fe (211). The crystal orientation dependence of TMR is considered to be originated from the crystal anisotropy of spin polarization.

(c) **Spin-valve-type junctions with an exchange-biased layer** In order to obtain a low and well defined switching magnetic field for a magnetic tunnel junction, the method of exchange biasing one of the magnetic layers using unidirectional magnetic anisotropy imposed by a thin antiferromagnetic layer was proposed. Junctions of this type have been studied first by Sato and Kobayashi [29, 30], Lu *et al.* [31] and Gallagher *et al.* [32]. They constructed

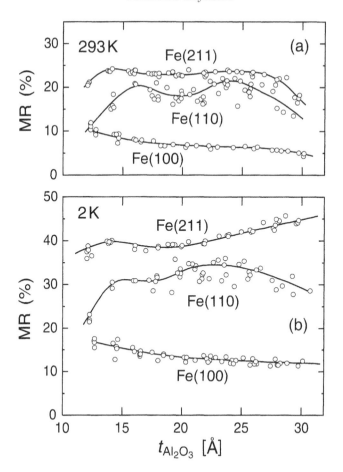

Figure 3.2: The TMR ratio of $Fe/Al_2O_3/Fe_{50}Co_{50}$ junctions at 293 K (a) and 2 K (b) as a function of the barrier thickness ($t_{Al_2O_3}$). The Fe electrode is a single crystal with (100), (110) and (211) planes.

the spin-valve-type junctions using permalloy and pure cobalt as the magnetic layers. The interesting point for these magnetoresistive properties is that the TMR ratio at room temperature is above 20%, which is larger than that without the antiferromagnetic layer. Furthermore, the switching magnetic field is a few tens of Oersteds. These results are very important for the application this type of junctions as electronic devices. After the reports for these junctions, various stackings with exchange-biased layer have been proposed. Especially, the junctions studied from the practical point of view are the spin-valve-type junctions and they are summarized in Table 3.3 [29-47]. As shown in Table 3.3, all the spin-valve-type junctions consist of many stacking layers.

Table 3.3: Spin-valve-type junctions reported

| Tunnel junction | Year | Authors |
| --- | --- | --- |
| • $(100)Si/Ni\text{-}Fe(171)/Co(33)/AlO_x(9\sim21)/Co(33)/NiFe(171)/FeMn(450)/Ni\text{-}Fe(86)$ | 1997 | Sato and Kobayashi [29, 30] |
| • $(100)Si/Pt(200)/Py(40)/MnFe(100)/Py(80)/Al_2O_3(10\sim30)/Co(80)/Pt(200)$ | 1997 | Gallagher et al. [32] |
| • $(100)Si/Ta(50)/Pt(200)/Py(40)/FeMn(100)/Pn(60)/Co(20Co)Al\text{-}O/Py(150)/Ta(200)$ | 1997 | Lu et al. [31] |
| • $(100)Si/Py(188)/Co(133)/Al_2O_3(8\sim16)/Co(26)/Py(188)/FeMn(450)/Ta(200)$ | 1997 | Kumagai et al. [33] |
| • $Si/NiFeCo(125)/Al_2O_3(25)/CoFe(70)/Ru(9)/CoFe(70)/FeMn(125)$ | 1998 | Tondra et al. [34] |
| • $glass/Ta(70)/NiFe(100)/Al_2O_3(\sim11)/CoFe(40)/MnRh(170)/Ta(30)$ | 1998 | Sousa et al. [35] |
| • $Si/SiO_2/Ti(50)/Pd(150)/Mn_{46}Fe_{54}(100)/Co_{84}Fe_{16}(30)/Al_2O_3(12\sim30)/Co_{84}Fe_{16}(80)/Pd(200)/Ti(50)$ | 1999 | Nowak et al. [36] |
| • $Si/SiO_2/Ta(50)/Al(250)/Ni_{60}Fe_{40}(40)/Mn_{54}Fe_{46}(100)/Co(40)/Ru(7)/Co(30)/AlO(\sim7)/$ $Ni_{60}Fe_{40}(75)/Al(250)/Ta(75)$ | 1999 | Parkin et al. [37] |
| • $glass/Ta(80)/Ni_{81}Fe_{19}(60)/Mn_{78}Rh_{22}(170)/Ni_{81}Fe_{19}(50)/Co_{90}Fe_{10}(20)/Al_2O_3/Co_{90}Fe_{10}(t)/Ta(30)$ | 1999 | Sun and Freitas [38] |
| • $SiO_2/Ta(35)/Ni_{80}Fe_{20}(30)/Fe_{50}Mn_{50}(20)/Ni_{80}Fe_{20}(25)/Co(15)/Al\text{-}O(15)/Co(40)/Ni_{80}Fe_{20}(100)/Ta(35)$ | 1999 | Gillies et al. [39] |
| • $AlTiC/Ta(50)/Cu(500)/Ta(50)/NiFe(100)/Co(20)/AlO_x(\sim10)/Co(30)/RuRhMn(100)/Ta(50)$ | 1999 | Shimazawa et al. [40] |
| • $SiO_2/NiFe(240)/Co\text{-}Fe(100)/AlO(16)/Co\text{-}Fe(100)/IrMn(500)/Al(100)$ | 1999 | Kikuchi et al. [41] |
| • $SiO_2/NiFe(188)/Co\text{-}Fe(39)/AlO(13)/Co\text{-}Fe(50)/IrMn(150)$ | 1999 | Sugawara et al. [42] |
| • $100Si/Si_3N_4(2000)/NiFeCo//Al_2O_3(\sim15)/CoFe/IrMn$ | 1999 | Wang et al. [43] |
| • $100Si/Si_3N_4(2000)/NiFeCo//Al_2O_3(\sim15)/CoFe/Ru/CoFe/InMn$ | 1999 | Wang et al [43] |
| • $SiO_2/Ta(30)/Py(20)/FeMn(100)/Py(50)/A\text{-}oxide/Py(100)/Ta(50)$ | 1999 | Matsuda et al. [44] |
| • $(111)Si/Cr(15)/Fe(60)/Cu(300)/Co(18)/Ru(8)/Co(30)/Al_2O_3/Co(10)/Fe(60)/Cu(100)/Cr(50)$ | 1999 | Tiusan et al. [45] |
| • $SiO_2/Ta(20)/Al(300)/Ta(30)/Py(30)/IrMn(150)/Co(50)/AlO(6\sim13)/Co(50)/Py(200)/Ta(100)$ | 1999 | our group [46] |
| • $SiO_2/Ta(30)/Cu(200)/Py(30)/IrMn(100)/Co_{75}Fe_{25}(40)/AlO(\sim8)/Co_{75}Fe_{25}(40)/Py(200)/Ta(50)$ | 1999 | our group [47] |

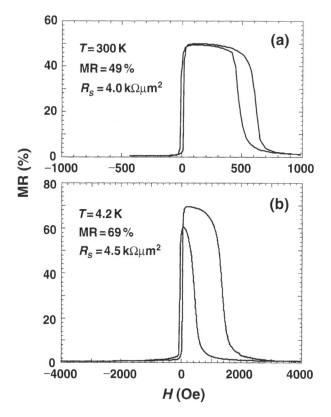

Figure 3.3: Typical MR curves measured at 300 K (a) and at 4.2 K (b) after annealing at 300 °C for spin-valve-type $SiO_2/Ta(50)/Py(30)/Cu(200)/Py(30)/IrMn(100)/Co_{75}Fe_{25}(40)/Al(8)$-O/$Co_{75}Fe_{25}(40)/Py(200)/Ta(50)$ junction. The value in the bracket is the layer thickness (Å).

The FeMn, MnRh, and IrMn are the typical antiferromagnetic layers and their Néel temperature is important related to their thermal stability.

Figure 3.3 shows a typical MR curve measured at room temperature and 4.2 K for $SiO_2/Ta(50)/Py(30)/Cu(200)$ /$Py(30)/IrMn(100)/Co_{75}$-$Fe_{25}(40)/Al$ ($\sim$8)-O/$Co_{75}$-$Fe_{25}(40)/Py(200)/Ta(50)$ spin-valve-type junction. The value in the bracket is the layer thickness (Å). In addition to the large valve of the TMR ratio at room temperature, the magnetoresistive curve exhibits a small hysteresis and it changes at small driving fields. These are the advantages for using the junction for magnetic read head in high-density magnetic recording and magnetic memory. Furthermore, the TMR ratio at 4.2 K is 69% and the highest one ever reported. According to the simple model of TMR ratio discussed by Julliere and Maekawa *et al.*, it can be expressed as TMR=$(R_{AP} -$

Table 3.4: Double or triple barrier layers junctions

| Tunnel junction | Year | Authors |
|---|---|---|
| • Ni/NiO/Co/NiO/Ni | 1997 | Ono *et al.* [48] |
| • MgO/Fe/Co$_{80}$Pt$_{20}$/SiO$_2$/Co$_{80}$Pt$_{20}$/SiO$_2$/Co$_{80}$Pt$_{20}$/Co$_9$Fe | 1997 | Inomata *et al.* [51] |
| • Si/Co/Al$_2$O$_3$/Co(20)/Al$_2$O$_3$/Ni$_{80}$Fe$_{20}$/Au | 1997 | Montaigne *et al.* [52] |
| • SiO$_2$/NiFe(190)/Co(40)/AlO($\sim$13)/Co(8)/AlO($\sim$13)/ Co(40)/NiFe(100)/FeMn(500)/Co(100) | 1998 | Fukumoto *et al.* [50] |
| • SiO$_2$/NiFe(190)/Co(40)/AlO($\sim$13)/Co(8)/AlO($\sim$13)/ Co(13)/AlO($\sim$13)/Co(300) | 1998 | Fukumoto *et al.* [50] |
| • SiO$_2$/Fe(60)/CoFe(30)/Al$_2$O$_3$(15)/CoPt(19)/ Al$_2$O$_3$(26)/CoFe(60)/Fe(120)/Au(980) | 1999 | Nakajima *et al.* [49] |

$R_P)/R_P = 2P_1P_2/(1 - P_1P_2)$, where $R_{AP}$ and $R_P$ are the tunnel resistance for antiparallel and parallel states of magnetization, respectively. $P_1$ and $P_2$ are the spin-polarizations for the first and second magnetic layers, respectively. In the present case, $P_1 = P_2$, and the value of $P_1$ ($P_2$) is about 50%. The calculated TMR ratio using this relation is approximately 67%, which is very near to the experimentally obtained value. In this experiment, an ideal value of TMR ratio is obtained.

**(d) Double and triple barrier layers junctions** The TMR study on the junctions with double or triple barrier layers attracts much attention from both fundamental physical and practical points of view. Some interesting physical phenomena, magneto-Coulomb oscillation [48], TMR oscillation with increasing applied voltage [49], and enhanced TMR ratio at Coulomb blockade (CB) regime [50] are already reported. By controlling the interface structure and fabricating a well designed junction, we are able to find a new physical phenomena. As will be described in Sec. 3.3, TMR ratio decreases rapidly with increasing applied voltage. Therefore, the improvement of bias dependence is important for the application of TMR junctions as electronic devices. By making the double barrier junction, enhanced TMR at high bias voltage is expected [51, 52]. Table 3.4 summarizes the double or triple barrier layers junctions experimentally studied [48-52].

**(e) Other tunnel magnetoresistance materials** Besides the F/I/F junctions, TMR effect has been also studied for granular materials. Especially, a relatively large value of TMR ratio was also found for Co-Al-O granular [53] nearly at the same time when a large TMR ratio at room temperature was obtained in typical layered junctions. In the last 5 years, various kinds of TMR study were reported for Co-Al-O [53-57], Fe-Hf-O [58], Fe-Mg-F [59], Co-SiO$_2$ [60], Fe-SiO$_2$ [61] and Ni-SiO$_2$ [62] granulars. The theoretical explanation of the temperature and bias dependence of TMR ratio and tunnel resistance, enhancement of TMR ratio in the Coulomb blockade regime have been discussed and described in Chapter 4 of this book. Physically, the transport properties of the granular is interesting, especially related to the charging effect of small granular particles. In general, the TMR effect of the granular materials occurs at high magnetic fields and the value is smaller than that of junctions. As granular materials have a low magnetic sensitivity, therefore, granular tunnel magnetoresistance has no possibility for practical use unless they have an increase of magnetic sensitivity.

Usually, in F/I/F junctions F is the ferromagnet with $3d$ transition metals such as Fe, Co, Ni and their alloys. However, $La_{1-x}Sr_xMnO_3$ and $SrTiO_3$ can also be chosen as F and I, respectively, and the junction is expressed as LSMO/STO/LSMO. For such junction [63-65] and layered perovskites $La_{2-2x}Sr_{1+2x}Mn_2O_7$ [66], the TMR properties were studied extensively. A large TMR ratio has been observed at lower temperatures. These studies are motivated due to the large spin polarizations (100%). This extremely large TMR ratio has been named colossal magnetoresistance (CMR). More detailed explanation are given in another book of the same monograph series.

## 3.3 Temperature and applied voltage dependence of TMR ratio

In spite of the high sensitivity of magnetic field and large TMR ratio, TMR effect has been found to show a strong temperature and applied voltage dependence. From both physical and practical points of view, it is important to understand the temperature and applied voltage dependence of TMR ratio and their origin. In this section, we briefly discuss these two phenomena.

(a) **Temperature dependence of TMR ratio**  Inoue proposed the temperature dependence of the TMR ratio in consideration of the magnetic impurity assisted tunneling [67]. The spin flip scattering due to the magnetic impurity causes the rapid decrease of TMR ratio in a low temperature range. On the other hand, Zhang *et al.* [68] discussed the decrease of TMR ratio with increasing temperature and applied voltage by spin excitations localized at interface. Generally speaking, spin flip tunneling due to the magnons decreases the TMR ratio of the junction. We show here a comparison between our experimental data and numerical result calculated by the models described above.

We assume that the total tunnel conductance consists of the conductance of elastic tunnel process $(G_n)$, the magnetic impurity assisted tunnel process $(G_j)$ and the magnon assisted tunnel process $(G_m)$. Then, the total tunnel conductance is expressed as [69, 70],

$$G = G_n + G_j + G_m, \tag{3.1}$$

where

$$G_n = G_0 T_n^2 B_{P(AP)},$$
$$G_j = 2G_0 T_j^2 \left[ \langle m^2 \rangle B_{P(AP)} + \langle l_+^2 \rangle F(\beta\Delta) B_{AP(P)} \right],$$
$$G_m = 2G_0 T_m^2 \left[ S^2 B_{P(AP)} + (Sk_B T/E_m) \ln(k_B T/E_c) B_{AP(P)} \right],$$
$$B_P = D_{1+} D_{3+} + D_{1-} D_{3-},$$
$$B_{AP} = 2D_{1+} D_{3-},$$
$$\langle m^2 \rangle = S(S+1) - \langle m \rangle^2 \coth(\beta\Delta),$$
$$\langle l_+^2 \rangle = \langle m \rangle \left[ \coth(\beta\Delta/2) - 1 \right],$$
$$F(\beta\Delta) = \beta\Delta / \left[ 1 - \exp(\beta\Delta) \right],$$
$$\beta = 1/k_B T,$$
$$\Delta = g\mu_B h_{\text{eff}},$$

and

$$E_m = 3k_B T_c / (S+1).$$

Here, $T_n^2$, $T_j^2$, and $T_m^2$ are the tunnel probabilities for normal, impurity and magnon assisted process, respectively. $G_0$ is a constant, $D_{i+}$ and $D_{i-}$ are the densities of states for majority and minority electrons in the $i$-th layer, $S$ is a spin operator, $g$ is a g factor, $\mu_B$ is the Bohr magneton, $h_{\text{eff}}$ is an effective internal field, $\langle m \rangle$ is the Brillouin function and $E_c$ is a cut off energy of magnon.

Figure 3.4 shows an example of temperature dependence of conductance for parallel and antiparallel configurations. In the figure, both experimental data and best fitted curve using Eq. (3.1) are described. The fitted curves are very close to the experimental data. Therefore, we evaluated the temperature dependence of TMR ratio by using the fitted curve. Figure 3.5 shows the temperature dependence of TMR ratio for both experiment and calculation [71]. The calculation reproduces well the experimentally obtained data. In the calculation, we assumed that $S = 3/2$, $|T_m|^2/|T_n|^2 = 0.2$, and obtained the fitting parameters, $P$, $|T_j|^2/|T_n|^2$, $E_c$, $h_{\text{eff}}$ and $T_c$. According to Zhang *et al.* [68], $|T_m|^2/|T_n|^2 = 0.2$ is reasonable because $|T_n|^2$ should be 1 to 2 orders of magnitude larger than $|T_m|^2$. In the analysis with many fitting parameters, it is important to check whether their values are reasonable or not. $P$ is roughly constant between 0.2 and 0.3 above 100 Å $d_{\text{Co}}$ layer. The value $T_j^2/T_n^2 \leq 0.1$ is nearly the same as reported one by analyzing the magnetic field dependence of zero bias anomaly. The value of $h_{\text{eff}} \approx 200\,\text{kOe}$ is exchreasonable when compared with the internal field of a ferromagnet. The $T_c$ is around 300 K which is much smaller than that of bulk value. The value of $T_c$ of Co at

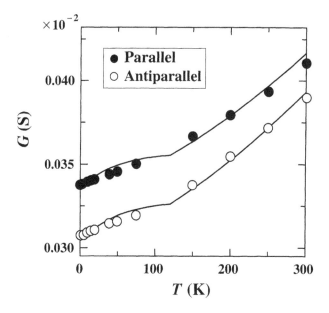

Figure 3.4: Temperature dependence of the conductance for the Al/Co(20 Å)/Al-oxide/Co junction. The magnetization of first and third layer are parallel (•) and antiparallel (○) statement. Solid curves are fitted results by using Eq. (3.1).

interface may decrease because of the decrease of the number of surrounding Co. The value of $E_c$ is between 100 K and 150 K which may be reasonable according to some past literature.

In order to obtain a low and well defined switching field of magnetic tunnel junction and to improve the temperature dependence of TMR ratio, we have investigated the spin-valve-type junctions with exchange-biasing one of the magnetic layers using unidirectional magnetic anisotropy imposed by a thin antiferromagnetic FeMn [72] or IrMn [73] layer. These junctions were annealed at 250 °C for 1 hour under a magnetic field. The switching fields of the free layers in the FeMn and IrMn exchange-biased junctions were 3 Oe and 5 Oe, respectively. Figure 3.6 shows TMR ratio and resistance at saturation as a function of temperature [74]. It is seen that the decrease of TMR ratio with increasing temperature is further improved by making a spin-valve-type junctions and successive heat treatment. Especially, the TMR ratio for the junction with IrMn exchange biased layer keeps above 20% even up to above 150 °C. This is a great advantage for the practical uses of the tunnel junction. Also, these temperature dependence is well explained by the magnon effect.

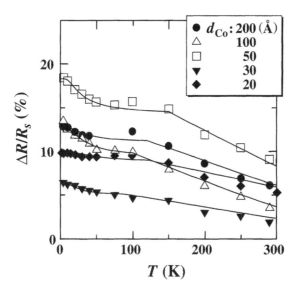

Figure 3.5: Temperature dependence of TMR ratio for Al/Co ($d_{Co}$Å)/Al-oxide/Co junctions.

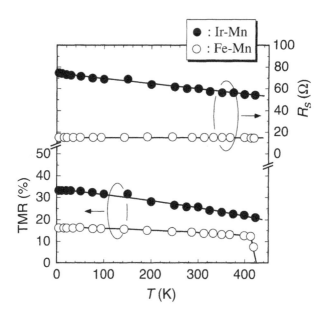

Figure 3.6: Temperature dependence of tunnel resistance and TMR ratio for spin-valve-type NiFe/Co/Al$_2$O$_3$/NiFe/IrMn (FeMn) junctions.

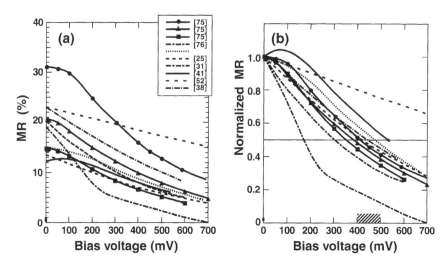

Figure 3.7: Applied voltage dependence of TMR ratio for various junctions reported.

**(b) Bias dependence of TMR ratio** One of the shortcomings of MRAM, is the well-known decrease of TMR ratio with increasing bias voltage. Figures 3.7(a) and (b) show the TMR ratio and normalized TMR ratio, respectively as a function of applied voltage. The TMR ratio can be expressed through the voltage $V_{1/2}$ at which the zero bias TMR value is halved. In the early stage of TMR study, $V_{1/2}$ was between 200 mV and 300 mV. As can be seen in Fig. 3.7(b) it increases up to about 500 mV from the recent data. This may be due to the improvement of oxidization process for insulator. It is very important, not to under or over oxidize the insulator layer. However, the exact reason is still unclear. Some part of decay is reduced in high quality junctions, but part of it seems to be due to some intrinsic mechanism such as excitation of magnons at the metal-barrier interfaces [68].

In order to suppress such decrease, tunnel junction with two or more barriers was introduced. It is expected that the $V_{1/2}$ of the two barriers junction becomes twice as that of single barrier junction. However, the decay of the TMR ratio of double barrier junction with bias voltage is significantly slower than expected from two independent junctions in series. Figure 3.8 shows an example of the comparison between experiment and calculation of the bias dependence of TMR ratio. The solid lines in the figure are the best fitted results. Since the fitting parameters such as $T_c$, cut off energy of magnon are consistent with those obtained from the analysis of temperature dependence of TMR ratio, magnon excitation is one of the possible origins of the TMR decay with increasing applied field.

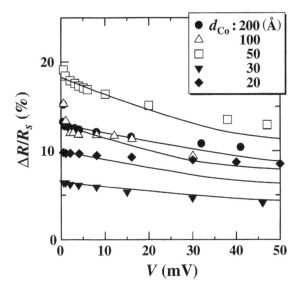

Figure 3.8: Applied voltage dependence of TMR ratio for Al/Co ($d_{Co}$Å)/Al-oxide/Co junctions.

## 3.4 Local transport properties

In order to make the tunnel junction usable as magnetic read head and random access memory, junction resistance must be accurately controlled and the fluctuation of the resistance should be reduced as much as possible. Therefore, characterization of local transport properties is required. A scanning probe microscope (SPM) system including an atomic force microscope (AFM) and scanning tunnel microscope (STM) is one of the most powerful instruments for achieving the characterization of electrical properties on a nanometer scale.

A conducting AFM, which is equipped with a conducting cantilever, can simultaneously provide topographical and electrical images, and has a nanometer-scale resolution. We used this method to evaluate the insulating barrier [77-79].

The triangular cantilever was made of silicon nitride and both sides were coated with about 400 Å thick Au. The topographical image was obtained by an ordinary contact-mode AFM. Simultaneously, a bias voltage was applied between the substrate and the tip and the current was mapped, resulting in an electrical (current) image.

Figure 3.9(a) and (b) show the topographical and current images measured simultaneously, respectively. The scanning area was $1500 \times 1500$ Å$^2$ and the bias voltage was 6 V. The average roughness evaluated from the topographical

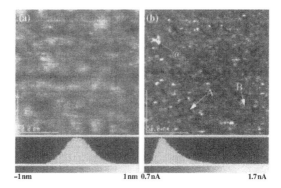

Figure 3.9:    (a)    The    topographical    image,    (b)    electrical    image    for
$Ni_{80}Fe_{20}$/Co/Al-oxide junction.

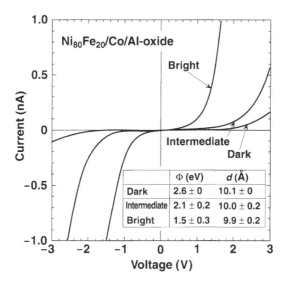

Figure 3.10: The typical *I-V* characteristic of dark, bright and intermediate regions of the electrical image.

AFM image was 2 Å. In the electrical image, the bright areas indicate a high tunneling current. It should be noted that there is no strong correlation between topographical and current images. The average size of the topographically high (bright) region is around 200 Å, while that of the electrically high region is only a few 10 Å. Furthermore, we can observe some extensive bright regions, which may be hot spots that cause the dielectric breakdown of the junction.

**Ta/Py/Pt/Py/IrMn/Co$_{75}$Fe$_{25}$/Al(8)-O   45 sec.**

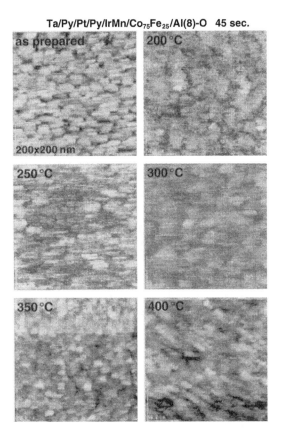

Figure 3.11: Current images for the junction annealed at various temperatures.

In order to discuss qualitatively the quality of the barrier, the *I-V* curves of some local points of the junction were measured. Figure 3.10 shows three typical examples of the *I-V* curves representative of the dark, bright and intermediate regions of the current image. These curves were fitted into the Simmons relation and the average barrier height $\Phi$ and the thickness $d$ are obtained (shown at the inset of Fig. 3.10). The barrier thickness is nearly the same in all locations, while the barrier height changes from place to place. This result indicates that the contrast in the current image provides information about the distribution of a local barrier height. This distribution may be due to a lack of oxygen on an atomic level in comparison with the Al$_2$O$_3$ composition.

In the following we will show that the method of AFM with current is very useful to investigate the quality of the insulator. Figure 3.11 shows the current image for the junction annealed at various temperatures. We can

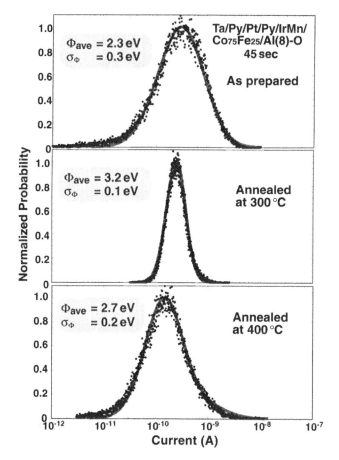

Figure 3.12: Histograms of current for the junction annealed at various temperatures.

clearly see that the current image become very homogeneous and smooth after annealing at around 300 °C. As described above, the contrast of current image provides a distribution of barrier height. Figure 3.12 shows the histogram of the current obtained from the current image shown in Fig. 3.11. In order to analyze these data quantitatively, current probability was fitted numerically with the following Gaussian distribution function [77],

$$P(\Phi_{ave}, \sigma_\Phi) = \frac{1}{\sqrt{2\pi}\sigma_\Phi} \exp\left[-\frac{(\Phi(J) - \Phi_{ave})^2}{2\sigma_\Phi^2}\right], \qquad (3.2)$$

Here, the current at parallel state of magnetization $J_p$ is given by [78, 80, 81],

$$J_p(\Phi) = a\frac{V^2}{\Phi d^2} \exp\left(b\frac{\Phi^{\frac{3}{2}}d}{V}\right), \tag{3.3}$$

where $a$ and $b$ are given by

$$a = A_{\text{eff}}\frac{e^3 m_0}{8\pi h m_{\text{eff}}} \cdot \frac{\beta^2}{t(E)^2}, \tag{3.4}$$

$$b = -\frac{8\pi(2m_{\text{eff}})^{\frac{1}{2}}}{3he} \cdot \frac{V(E)}{\beta}, \tag{3.5}$$

where $\Phi_{ave}$ and $\sigma_\Phi$ are the average and deviation values of barrier height $\Phi$, respectively. $V$ and $d$ in Eq. (3.3) are the applied voltage and barrier width, respectively. $A_{\text{eff}}$ in Eq. (3.4) is the effective area between cantilever and sample. $m_0$, $m_{\text{eff}}$, $e$ and $h$ are the mass of free electron, effective mass of tunnel electron, charge of electron and Planck's constant, respectively. The values of $\beta$, $t(E)$ and $V(E)$ are the values depending on the shape of cantilever. The solid curves in Fig. 3.12 are the best fitted result using Eqs. (3.2)-(3.5). In the figure, the average barrier height ($\Phi_{ave}$) and its standard deviation ($\sigma_\Phi$) are also indicated. The TMR ratio of the junction is considered to be the average of local TMR values and can be expressed as,

$$TMR_{\text{total}} = \frac{\sum_\Phi[R_P(\Phi) - R_{AP}(\Phi)] \cdot P(\Phi_{ave}, \sigma_\Phi)}{\sum_\Phi R_P(\Phi)P(\Phi_{ave}, \sigma_\Phi)}$$
$$= \frac{\sum_\Phi[1/J_P(\Phi) - 1/J_{AP}(\Phi)] \cdot P(\Phi_{ave}, \sigma_\Phi)}{\sum_\Phi 1/J_P(\Phi)P(\Phi_{ave}, \sigma_\Phi)}, \tag{3.6}$$

where $J_{AP}(\Phi) = J_P(\Phi)[1-TMR_{\text{cal}}(\Phi)]$. Therefore, if we know the $TMR_{\text{cal}}(\Phi)$, TMR ratio can be evaluated by Eq. (3.6). We used the values of $J_P(\Phi)$ and $TMR_{\text{cal}}(\Phi)$ reported by Itoh *et al.* [82], respectively. Figure 3.13 shows the TMR ratio as a function of annealing temperature for both experimentally obtained and evaluated values [83]. It is clearly seen that the increase of TMR with annealing can be well explained by taking into account of the change of distribution of the tunnel barrier height.

## 3.5 Inelastic Electron Tunneling Spectroscopy

Inelastic Electron Tunneling (IET) Spectroscopy was proposed as a unique method to investigate an electron state of an interface of ferromagnet/insulator. We applied this method to analyze a spin dependent tunneling process [84-89]. Figure 3.14 shows the three types of models for the interlayer structure and their corresponding $dI/dV - V$ curves and IET spectra. If there is a

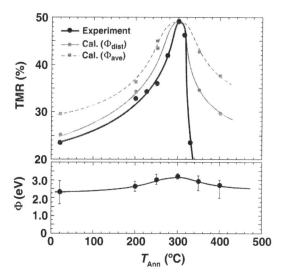

Figure 3.13: TMR ratio as a function of annealing temperature for both experimentally obtained and calculated using the histograms of current.

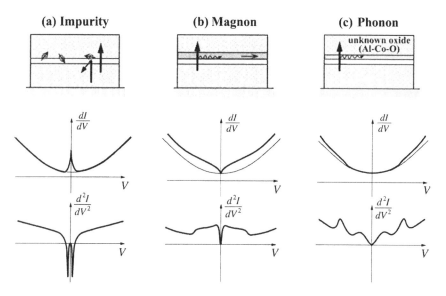

Figure 3.14: The schematic illustrations of three type models for the interlayer structure and corresponding $dI/dV - V$ curves and IET spectra. —— : normal conductance, — : expected conductance.

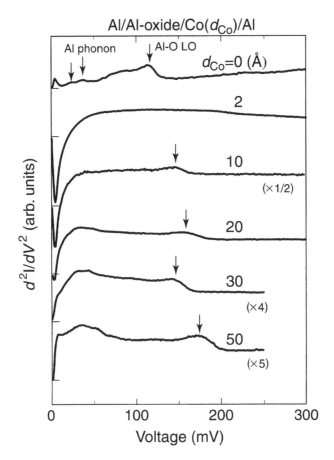

Figure 3.15: IET spectra of $Al/Al_2O_3/Co(d_{Co})/Al$ junctions.

paramagnetic impurity at the interface between an electrode and insulator (Fig. 3.14(a)), an inelastic electron tunneling process would occur in addition to an elastic tunneling process [84, 85]. In this case, the conductance should increase around zero-bias voltage and IET spectrum gives a strong negative peak. When the magnetic layer thickness becomes thick enough to form a continuous ferromagnetic layer (see Fig. 3.14(b)), a magnon-assisted electron tunneling process would be added [68]. In this case, IET spectrum should show some positive peaks corresponding to the magnon density of states. In the case of the presence of a phonon assisted inelastic tunneling process due to an unknown oxide (see Fig. 3.14(c)), we would observe a similar IET spectrum corresponding to the phonon density of state. Figure 3.15 shows experimentally obtained IET spectra for $Al/Al_2O_3/Co$ ($d_{Co}$)$/Al$, $d_{Co} = 0 \sim 50$ Å junctions.

These junctions were fabricated on glass substrates by rf magnetron sputtering. The base pressure was below $8 \times 10^{-5}$ Pa and the sputtering process was carried out in an atmosphere of 0.2 (0.6) Pa Ar with an electrical input of 4.4 (1.1) W/cm$^2$ for Al (Co). The bottom Al electrode with 0.5 mm width was prepared using a mask. The Al was naturally oxidized in air at 60 °C for 24 hours. The top Co/Al electrode with 0.5 mm width was formed by the subsequent deposition over the insulator to form a cross-patterned junction. IET spectra was measured by a modulation method using a hand made electrical circuit [84]. In addition to a dc voltage sweep, a modulation voltage of 4 mV with frequency of 5 kHz was applied to the junctions. We measured the voltage at the first and second harmonic frequency by a lock-in amplifier. As can be seen in the figure, a strong negative peak at about 4 mV corresponding to the model (a) in Fig. 3.14 is observed for $d_{Co} = 2$ Å. The peak intensity tends to decrease with increasing $d_{Co}$ and disappears at $d_{Co} = 30$ Å. The peaks at 27 and 33 mV for the junction of $d_{Co} = 0$ are assigned to be Al-phonon and that at 120 mV to be longitudinal optical (LO) mode of Al-O bonds. Such phonon spectra become obscure for the junction with $d_{Co} = 2$ Å. A broad peak (arrows in the figure) is observed for the junction with $d_{Co} \geq 10$ Å. The peak position shifts slightly to higher voltage and corresponds to the model (b) in Fig. 3.14. However, it is presently difficult to distinguish the magnon and phonon effects.

## 3.6 Summary

One of the major reasons for the TMR effect has attracted so much attention by many researchers is that it processes a high potential for the applications of magnetic read heads in hard disk drives and as well as magnetoresistive (or magnetic) random access memory. Both hard disk drives and magnetic memories are large industries and are expected to grow larger and faster in our global communication networks. Moreover, recently we sometime use or hear the words, "spin electronics" and "magnetoelectronics" [90], which have the same meaning. Nanoscale magnetism and transport play an important role in the field of magnetic thin films is one of the big topics in the nanoscale magnetism and transport. So, the TMR effect is a very attractive research subject for both fundamental and applied physics. Here I briefly describe mainly from the materials point of view of the recent development for the study related to magnetic memory and read head applications. More detailed explanation is given in Chapter 5 of this book.

Figure 3.16 summarizes the TMR ratio and tunnel resistance reported by various groups. In the figure, the region of Target (Head) is the values of TMR ratio and tunnel resistance required for magnetic read head in hard disk drives. The value of resistance per 1 $\mu$m$^2$ is roughly the same value as that currently

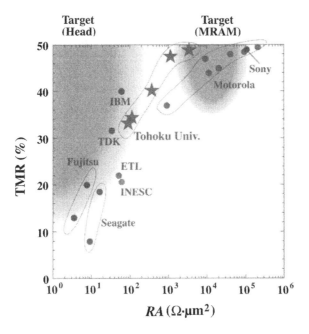

Figure 3.16: TMR ratio and tunnel resistance reported by various groups. The shaded regions of target (head) and (MRAM) are the values for reading head and magnetic random access memory, respectively.

producing GMR head. Since the MR ratio currently produced is less than 10%, about two times MR ratio is large enough for practical use. The region of Target (MRAM) is the values of those required for magnetic (or magnetoresistive) random access memory. Please note the these values are only estimations [91]. From the materials point of view, the transport properties required for MRAM is already satisfied. Recent study for magnetoresistive random access memory is related to memory cell design [92] and circuit architecture [37, 93-98]. Furthermore, reproducibility of uniform tunnel resistance and reduction of switching field are actively studied. The advantage of MRAM over the commercially available high density memory such as DRAM, SRAM and FLASH memories is that MRAM is non-volatile. To be commercially viable, MRAM technology must meet the following criteria: comparable memory density, scalable for future generations, operate at low voltages, competitive read and write speeds, and have low power consumption.

On the other hand, the tunnel resistance required for read head should be two or three order smaller than that for MRAM. Therefore, the production of uniform and thin insulating layer is very important. Several companies are very active in this field. Recent development for MRAM is explained in Chapter 5 of this book.

# References

[1] L. M. Falicov, D. T. Pierce, S. D. Bader, R. Gronsky, K. B. Hathaway, H. J. Hopster, D. N. Lambeth, S. S. P. Parkin, G. Prinz, M. Salamon, Ivan K. Schuller and R. H. Victora, J. Mater. Res. **5**, 1299 (1990).

[2] S. D. Bader, Proceedings of The IEEE **78**, 909 (1990).

[3] U. Gradmann, J. Magn. Magn. Mater. **100**, 481 (1991).

[4] T. Shinjo, Surface Science Reports, **12**, 49 (1991).

[5] H. C. Siegmann, J. Phys. Condens. Matter, **4**, 8395 (1992).

[6] T. Miyazaki, Electrochemical Technology, 279 (1996).

[7] F. J. Himpsel, J. E. Ortega, G. J. Mankey and R. F. Willis, Advances in Physics, **47**, 511 (1998).

[8] M. Julliere, Phys. Lett. **54A**, 225 (1975).

[9] S. Maekawa and U. Gäfvert, IEEE Trans. Magn. MAG-**18**, 707 (1982).

[10] Y. Suezawa and Y. Gondo, Proc. Int. Symp. Physics of Magnetic Materials, Sendai, MAG-**18**, 303 (1987).

[11] R. Nakatani and M. Kitada, J. Mater. Sci. Lett. **10**, 827 (1991).

[12] T. Miyazaki, T. Yaoi and S. Ishio, J. Magn. Magn. Mater. **109**, L7 (1991).

[13] T. Yaoi, S. Ishio and T. Miyazaki, J. Magn. Magn. Mater. **98**, 430 (1993).

[14] T. Miyazaki and N. Tezuka, J. Magn. Magn. Mater. **139**, L231 (1995).

[15] J. S. Moodera, L. R. Kinder, T. M. Wong and R. Meservey, Phys. Rev. Lett. **74**, 3273 (1995).

[16] J. Nowak and J. Rauluszkiewicz, J. Magn. Magn. Mater. **109**, 79 (1992).

[17] Y. Suezawa, F. Takahashi and Y. Gondo, Jpn. J. Appl. Phys. **31**, L1415 (1992).

[18] T. S. Plaskett, P. P. Freitas, N. P. Barradas, M. F. da Silva and J. C. Sores, J. Appl. Phys. **76**, 6104 (1994).

[19] N. Tezuka and T. Miyazaki, J. Appl. Phys. **79**, 6262 (1996).

[20] J. S. Moodera and L. R. Kinder, J. Appl. Phys. **79**, 4724 (1996).

[21] C. L. Platt, B. Dieny and A. E. Berkowitz, J. Appl. Phys. **81**, 5523 (1997).

[22] R. B. Beech, J. Anderson, J. Daughton, B. A. Everitt and D. Wang, IEEE Trans. Magn. **33**, 4713 (1996).

[23] C. T. Tanaka, J. Nowak and J. S. Moodera, J. Appl. Phys. **81**, 5515 (1997).

[24] J. S. Moodera, E. F. Gallagher, K. Robinson and J. Nowak, Appl. Phys. Lett. **70**, 3050 (1997).

[25] H. Tsuge and T. Mitsuzuka, Appl. Phys. Lett. **71**, 3296 (1997).

[26] W. Oepts, H. J. Verhagen and W. J. M. de Jonge, Appl. Phys. Lett. **73**, 2363 (1998).

[27] J. Zhang and R. M. White, J. Appl. Phys. **83**, 6512 (1998).

[28] S. Yuasa, T. Sato, Y. Suzuki, E. Tamura, T. Katayama and K. Ando, Europhys. Lett. **52**, 344 (2000).

[29] M. Sato and K. Kobayashi, Jpn. J. Appl. Phys. **36**, L200 (1997).

[30] M. Sato and K. Kobayashi, J. Magn. Soc. Jpn. **21**, 489 (1997).

[31] Yu Lu, R. A. Altman, A. Marley, S. A. Rishton, P. L. Trouilloud, G. Xiao, W. J. Gallagher and S. S. P. Parkin, Appl. Phys. Lett. **70**, 2610 (1997).

[32] W. J. Gallagher, S. S. P. Parkin, Yu Lu, X. Y. Bian, A. Marley, K. P. Roche, R. A. Altman, S. A. Rishton, C. Jahnes, T. M. Shaw and Gang Xiao, J. Appl. Phys. **81**, 3741 (1997).

[33] S. Kumagai, N. Tezuka and T. Miyazaki, Jpn. J. Appl. Phys. **36**, L1498 (1997).

[34] M. Tondra, J. M. Daughton, D. Wang, R. S. Beech, A. Fink and J. A. Taylor, J. Appl. Phys. **83**, 6688 (1998).

[35] R. C. Sousa, J. J. Sun, V. Soares, P. P. Freitas, A. Kling, M. F. da Silva and J. C. Soares, Appl. Phys. Lett. **73**, 3288 (1998).

[36] E. R. Nowak, M. B. Weissman and S. S. P. Parkin, Appl. Phys. Lett. **74**, 600 (1999).

[37] S. S. P. Parkin, K. P. Roche, M. G. Samant, P. M. Rice, R. B. Beyers, R. E. Scheuerlein, E. J. O 'Sulivan, S. L. Brown, J. Bucchigano, D. W. Abraham, Yu Lu, M. Rooks, P. L. Trouilloud, R. A. Wanner and W. J. Gallagher, J. Appl. Phys. **85**, 5828 (1999).

[38] J. J. Sun and P. P. Freitas, J. Appl. Phys. **85**, 5264 (1999).

[39] M. F. Gillies, W. Oepts, A. E. T. Kuiper, R. Coehoorn, Y. Tamminga, J. H. M. Snijders and W. M. Arnold Bik, IEEE Trans. Magn. **35**, 2991 (1999).

[40] K. Shimazawa, N. Kasahara, Jijun San, S. Araki, H. Morita and M. Matsuzaki, J. Appl. Phys. **87**, 5194 (2000).

[41] M. Sato, H. Kikuchi and K. Kobayashi, IEEE Trans. Magn. **35**, 946 (1999).

[42] J. Sugawara, E. Nakashio, S. Kumagai, J. Honda, Y. Ikeda and T. Miyazaki, J. Magn. Soc. Jpn. **23**, 1281 (1999).

[43] D. Wang, M. Tondra, J. M. Daughton, C. Nordman and A. V. Pohm, J. Appl. Phys. **85**, 5255 (1999).

[44] K. Matsuda, A. Kamijo, T. Mitsuzuka and H. Tsuge, J. Appl. Phys. **85**, 5261 (1999).

[45] C. Tiusan, M. Hehn, K. ounadjela, Y. Heny, J. Hommet, C. Meny, H. van den Berg, L. Baer and R. Kinder, J. Appl. Phys. **85**, 5276 (1999).

[46] M. Kamijo, J. Murai, H. Kubota, Y. Ando, T. Miyazaki, C. Kim and O. Song, J. Magn. Soc. Jpn. **24**, 591 (2000).

[47] X. F. Han, T. Daibou, M. Kamijo, K. Yaoita, H. Kubota, Y. Ando and T. Miyazaki, Jpn. J. Appl. Phys. **39**, L439 (2000).

[48] K. Ono, H. Shimada and Y. Ootuka, J. Magn. Soc. Jpn. **66**, 1261 (1997).

[49] K. Nakajima, Y. Saito, S. Nakamura and K. Inomata, J. Magn. Soc. Jpn. **24**, 575 (2000).

[50] Y. Fukumoto, H. Kubota, Y. Ando and T. Miyazaki, Jpn. J. Appl. Phys. **38**, L932 (1999).

[51] K. Inomata, H. Ogiwara, Y. Saito, K. Yusu and K. Ichihara, Jpn. J. Appl. Phys. **36**, L1380 (1997).

[52] F. Montaigne, J. Nassar, A. Vaures, F. Nguyen Van Dau, F. Petroff, A. Schuhl and A. Fert, Appl. Phys. Lett. **73**, 2829 (1998).

[53] H. Fujimori, S. Mitani and S. Ohnuma, Materials Science and Engineering, B**31**, 219 (1995).

[54] H. Fujimori, S. Mitani and S. Ohnuma, J. Magn. Magn. Mater. **156**, 311 (1996).

[55] H. Fujimori, S. Mitani and S. Ohnuma, J. Magn. Magn. Mater. **165**, 141 (1997).

[56] T. Watabe, H. Kubota and T. Miyazaki, J. Magn. Soc. Jpn. **21**, 457 (1997).

[57] S. Mitani, H. Fujimori and S. Ohnuma, J. Magn. Magn. Mater. **177-181**, 919 (1998).

[58] Y. Hayakawa, N. Hasegawa, A. Makino, S. Mitani and H. Fujimori, J. Magn. Magn. Mater. **154**, 175 (1996).

[59] T. Furubayashi and I. Nakatani, J. Appl. Phys. **79**, 6258 (1996).

[60] S. Brazilai, Y. Goldstein, I. Balberg and J. S. Helman, Phys. Rev. B **23**, 1809 (1981).

[61] S. Honda, T. Okada and M. Nawate, J. Magn. Magn. Mater. **165**, 153 (1997).

[62] A. Milner, A. Gerber, B. Groisman, M. Karpovsky and A. Gladkikh, Phys. Rev. Lett. **76**, 475 (1996).

[63] Yu Lu, X. W. Li, G. Q. Gong, Gang Xiao, A. Gupta, P. Lecoeur, J. Z. Sun, Y. Y. Wang and V. P. Dravid, Phys. Rev. B **54**, R8357 (1996).

[64] M. Viret, M. Drouet, J. Nassar, J. P. Contour, C. Fermon and A. Fert, Europhys. Lett. **39**, 545 (1997).

[65] J. Nassar, M. Viret, M. Drouet, J. P. Contour, C. Fermon and A. Fert, Mat. Res. Soc. Symp. Proc. **494**, 231 (1998).

[66] T. Kimura, Y. Tomioka, H. Kuwahara, A. Asamitsu, M. Tamura and Y. Tokura, Science, **274**, 1698 (1996).

[67] J. Inoue, J. Phys. D : Appl. Phys. **31**, 643 (1998).

[68] S. Zhang, P. M. Levy, A. C. Marley and S. S. P. Parkin, Phys. Rev. Lett. **79**, 3744 (1997).

[69] J. Appelbaum, Phys. Rev. Lett. **17**, 91 (1966).

[70] P. W. Anderson, Phys. Rev. Lett. **17**, 95 (1966).

[71] M. Oogane, N. Tezuka and T. Miyazaki, J. Magn. Soc. Jpn. **23**, 1309 (1999).

[72] S. Kumagai, N. Tezuka and T. Miyazaki, Jpn. J. Appl. Phys. **36**, L1498 (1997).

[73] J. Sugawara *et al.*, (unpublished).

[74] T. Miyazaki, M. Oogane, Y. Fukumoto, S. Ootsuka, N. Tezuka, H. Kubota and Y. Ando, Final Report for the NEDO International Joint Research Grand Program 96MB1, S40 (1999).

[75] H. Kubota *et al.*, (unpublished).

[76] J. S. Moodera, J. Nowak and R. J. M. Veerdonk, Phys. Rev. Lett. **80**, 2941 (1998).

[77] Y. Ando, H. Kameda, H. Kubota and T. Miyazaki, Jpn. J. Appl. Phys. **38**, L737 (1999).

[78] Y. Ando, H. Kameda, H. Kubota and T. Miyazaki, J. Appl. Phys. **87**, 5206 (2000).

[79] Y. Ando, H. Kameda, M. Hayashi, H. Kubota and T. Miyazaki, J. Magn. Soc. Jpn. **24**, 611 (2000).

[80] R. H. Fowler and L. Nordheim, Proc. R. Soc. London Ser. A**119**, 173 (1928).

[81] S. J. O'shea, R. M. Atta, M. P. Murrell and M. E. Welland, J. Vac. Sci. Technol. B **13**, 1945 (1995).

[82] H. Itoh, A. Shibata, T. Kumazaki, J. Inoue and S. Maekawa, J. Phys. Soc. Jpn. **68**, 1632 (1999).

[83] Y. Ando, M. Hayashi, M. Kamijo, H. Kubota and T. Miyazaki, Abstract of International Symposium on Nanoscale Magnetism and Transport (IS-NMT 2000), p. 58 (2000).

[84] J. Murai, Y. Ando, N. Tezuka and T. Miyazaki, J. Magn. Soc. Jpn. **22**, 573 (1998).

[85] J. Murai, Y. Ando and T. Miyazaki, J. Magn. Soc. Jpn. **23**, 64 (1999).

[86] J. Murai, Y. Ando and T. Miyazaki, J. Magn. Soc. Jpn. **23**, 1325 (1999).

[87] J. Murai, Y. Ando, M. Kamijo, H. Kubota and T. Miyazaki, Jpn. J. Appl. Phys. **38**, L1106 (1999).

[88] J. Murai, Y. Ando, M. Kamijo, T. Daibou, H. Kubota T. Miyazaki, C. Kim and O. Song, J. Magn. Soc. Jpn. **24**, 615 (2000).

[89] Y. Ando, J. Murai, H. Kubota and T. Miyazaki, J. Appl. Phys. **87**, 5209 (2000).

[90] G. Prinz and K. Hathaway, Physics Today, April 24 (1995).

[91] K. Inomata, J. Magn. Soc. Jpn. **23**, 1826 (1999).

[92] J. M. Daughton, J. Appl. Phys. **81**, 3758 (1997).

[93] Z. Wang and Y. Nakamura, Jpn. J. Appl. Phys. **20**, 369 (1996).

[94] S. Tehrani, E. Chen, M. Durlam, M. DeHerrera, J. M. Slaughter and J. Shi, J. Appl. Phys. **85**, 5822 (1999).

[95] R. E. Scheuerlein, Proc. of Int'l NonVolatile Memory Technology Conference, p. 47 (1998).

[96] S. Tehrani, J. M. Slaughter, E. Chen, M. Durlam, J. Shi and M. DeHerrea, IEEE Trans. Magn. **35**, 2814 (1999).

[97] H. Boeve, C. Bruynseraede, J. Das, K. Dessein, G. Borghs Z, J. D. Boek, R. C. Sousa, L. V. Melo and P. P. Freitas, IEEE Trans. Magn. **35**, 2820 (1999).

[98] R. C. Sousa, P. P. Freitas, V. Chu and J. P. Conde, IEEE Trans. Magn. **35**, 2832 (1999).

# Chapter 4

# Theory of Tunnel Magnetoresistance

*Sadamichi Maekawa, Saburo Takahashi and Hiroshi Imamura*

## 4.1 Introduction

Since the discovery of giant magnetoresistance (GMR) in magnetic multilayers in 1988 [1], the spin-dependent transport phenomena in magnetic nanostructures has received much interest from both fundamental and technological points of view. Magnetic tunnel junctions consisting of two ferromagnetic layers (electrodes) separated by a thin insulating layer exhibit a large magnetoresistance more than 50% at room temperature, making promising candidate for the spin-electronic devices such as magnetic random access memories (MRAM). Recent advances in nanofabrication technique enables us to fabricate a new class of magnetic nanostructures such as granular films, single electron transistor, and atomic point contact, which exhibit novel magnetoresistive phenomena caused by the interplay between spin-dependent tunneling and the Coulomb blockade effect.

In this chapter, we briefly review theoretical understanding in physics of the tunnel magnetoresistance (TMR) with particular emphasis on recent developments in magnetic nanostructures. We begin with a brief introduction of theoretical models for describing spin-polarized tunneling and TMR in Sec. 4.2. We present three different models widely used in the study of TMR: tunnel Hamiltonian model in Sec. 4.2.1, free electron model in Sec. 4.2.2, and tight binding model in Sec. 4.2.3. The effect of the Coulomb blockade (CB) on the spin-dependent transport in various magnetic nanostructures are discussed in Sec. 4.3. After explaining the basic concept of the CB in magnetic nanostructures (Sec. 4.3.1), we investigate single-electron tunneling into a granule film using scanning tunneling microscope (STM) in Sec. 4.3.2. In Sec. 4.3.3 and Sec. 4.3.4, the effect of higher-order tunneling (cotunneling) on TMR in a double tunnel junction and a granular film is discussed, and a unified description for enhancement of TMR at low temperatures is presented. Spin accumulation in non-magnetic metals by injection of spin-polarized electrons from ferromagnets is another basic concept in magnetic nanostructures. The condition for spin accumulation in a non-magnetic metal sandwiched between two ferromagnetic electrodes is examined in Sec. 4.4.1. When a non-magnetic metal is a very small normal-metal, the spin-dependent CB occurs due to spin accumulation (Sec. 4.3.2). When a non-magnetic metal is a superconductor, a magneto-superconductive effect appears by strong competition between superconductivity and accumulated spins as shown in Sec. 4.4.3. Section 4.5 is devoted to other quantum effects on TMR: magnetic quantum point contact in Sec. 4.5.2 and Andreev reflection in Sec. 4.5.1.

# 4.2   Models of spin-dependent tunneling

Electron tunneling is a basic phenomenon in quantum mechanics by which electric current can pass from one electrode through a thin insulating barrier layer into a second electrode. In the early 1960s, the tunneling technique received considerable attention since Giaever [2] used it to obtain the density of states and the superconducting gap by measuring the tunnel conductance as a function of bias voltage, which are in excellent agreement with the Bardeen-Cooper-Schrieffer (BCS) theory.

In the early 1970s, Meservey, Tedrow, and coworkers [3] developed a spin-polarized electron tunneling technique which used special properties of the superconducting states to probe spin-dependent features of the electron density of states of magnetic metals. Applying this technique to various ferromagnets in ferromagnet/$Al_2O_3$/Al tunnel junctions, they showed that tunneling electrons from ferromagnets are spin-polarized, and obtained detailed information about the polarization of the conduction electrons near the Fermi level.

A few years after the discovery of the spin-polarized tunneling between a superconductor and a ferromagnet, tunneling between two ferromagnetic electrodes has been reported: Julliere [4] formed Fe/Ge/Co junctions and measured the tunnel conductance and showed that the magnetoresistance (TMR) depends on the relative orientation of the magnetic moments. Maekawa and Gäfvart [5] demonstrated a strong correlation between the tunnel conductance and magnetization process in Ni/NiO/ferromagnet junctions with Ni, Fe, or Co as the counter electrode. However, the value of measured TMR's remained to be very small at room temperature until the middle of 1990s. In 1995, a breakthrough for large TMR was brought about by two groups led by Miyazaki [6] and Moodera [7] by the advent of superior fabrication methods for magnetic tunnel junctions.

In the early work of the spin-dependent tunneling, the theoretical interpretation is based upon a simple model that the spin is conserved in the tunneling process and that the conductance of each spin direction is proportional to the densities of states of that spin in each electrode. In this model, the tunnel current is larger when the magnetic moments of the two electrodes are parallel than when they are antiparallel, which explains successfully strong dependence of tunneling current on the relative orientation of the magnetic moments of the ferromagnetic electrodes. However, the model does not explain the experimental results that the magnetic tunnel junctions exhibit rather large variations of bias and temperature dependencies in the TMR.

In 1989, Slonczewski suggested a different approach to describe the spin-dependent tunneling in magnetic tunnel junctions. By extending the free electron model of tunneling in non-magnetic tunnel junctions [8] to magnetic ones, Slonczewski found that the polarization of tunneling electrons depends not only on the electronic density of states of ferromagnets but also on the height of the tunnel barriers [9]. In this approach, the electrodes and the insulating barrier are treated as a single quantum-mechanical system and the wave functions of up and down electrons are constructed by solving the Schrödinger equation in the whole system. While the free-electron model captures some essential features, their predictions for TMR are quantitatively unreliable because the lattice structure of the electrodes and the variation of the band structure near the insulating barrier are overlooked [10].

More realistic and reliable descriptions for TMR are provided by the tight binding model which allows one to distinguish electronic structures at interfaces from that in the bulk and to study the effect of the interface roughness. It is shown that the conductance of the tight binding method reduces to the usual expression for the conductance obtained in the classical theory of tunneling when the electron hopping between the electrodes is weak enough and the coherence across the barrier is completely lost.

## 4.2.1 Tunnel Hamiltonian

When two metals are separated by a thin insulating barrier as shown in Fig. 4.1, there is a nonzero probability of charge transfer by quantum-mechanical tunneling of electrons between two metals through the insulating barrier. The basic idea of the tunnel Hamiltonian model is to write the Hamiltonian of the junction system as the sum of three terms:

$$\mathcal{H} = \mathcal{H}_L + \mathcal{H}_R + \mathcal{H}_T, \tag{4.1}$$

where $\mathcal{H}_L$ and $\mathcal{H}_R$ are the Hamiltonians of the left and right electrodes, respectively, and tunneling process is described by the tunnel Hamiltonian $\mathcal{H}_T$. The tunneling probability decreases exponentially with the thickness of the barrier and it depends on the details of the insulating material, but these aspects are absorbed in the phenomenological tunneling matrix elements $\hat{T}_{\mathrm{pk}}^{\sigma}$. Thus we

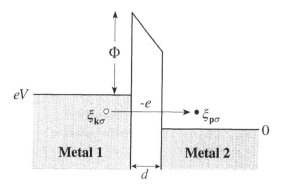

Figure 4.1: (a) Electron tunneling between two normal metals. The arrow indicates the electron with spin $\sigma$ transferred through the oxide barrier.

write the effective tunnel Hamiltonian of the form

$$\mathcal{H}_T = \sum_{\mathbf{k}\mathbf{p}\sigma} \hat{T}^\sigma_{\mathbf{p}\mathbf{k}} a^\dagger_{\mathbf{p}\sigma} a_{\mathbf{k}\sigma} + h.c., \qquad (4.2)$$

where $a_{\mathbf{k}\sigma}$ is the annihilation operator of an electron with wave vector $\mathbf{k}$ and spin $\sigma$ on the left electrode, $a^\dagger_{\mathbf{p}\sigma}$ the creation operator of an electron with wave vector $\mathbf{p}$ and spin $\sigma$ on the right electrode, and no spin flip is assumed in the tunneling process.

When the bias voltage $V$ is applied to the tunnel junction, the tunneling rate $\overrightarrow{\Gamma}_\sigma(V)$ at which electrons with spin $\sigma$ are transferred from the left to the right electrodes is calculated by Fermi's golden rule:

$$\overrightarrow{\Gamma}_\sigma(V) = \frac{2\pi}{\hbar} \sum_{\mathbf{k},\mathbf{p},\sigma} |\hat{T}^\sigma_{\mathbf{p}\mathbf{k}}|^2 f(\xi_{\mathbf{k}\sigma}) \left[1 - f(\xi_{\mathbf{p}\sigma})\right] \delta(\xi_{\mathbf{k}\sigma} - \xi_{\mathbf{p}\sigma} + eV), \qquad (4.3)$$

where $\xi_{\mathbf{k}}$ and $\xi_{\mathbf{p}}$ are one-electron energies measured from the Fermi levels and $f(\xi_{\mathbf{k}\sigma}) = \langle a^\dagger_{\mathbf{k}\sigma} a_{\mathbf{k}\sigma} \rangle = 1/\left[\exp(\xi_{\mathbf{k}\sigma}/k_B T) + 1\right]$ is the Fermi distribution function. The steady-state current through the junction is determined by the difference between the forward and backward tunneling rates:

$$I_\sigma(V) = e \left[\overrightarrow{\Gamma}_\sigma(V) - \overleftarrow{\Gamma}_\sigma(V)\right], \qquad (4.4)$$

where $\overleftarrow{\Gamma}_\sigma(V)$ is the tunneling rate at which electrons with spin $\sigma$ are transferred from the right to the left electrodes and is related with $\overrightarrow{\Gamma}_\sigma(V)$ by $\overleftarrow{\Gamma}_\sigma(V) = \overrightarrow{\Gamma}_\sigma(-V)$. Substituting Eq. (4.3) into Eq. (4.4), the tunnel current $I_\sigma$ for the spin channel $\sigma$ becomes

$$I_\sigma(V) = \frac{2\pi e}{\hbar} \langle |\hat{T}^\sigma_{\mathbf{p}\mathbf{k}}|^2 \rangle \int_{-\infty}^\infty \mathcal{D}_{1\sigma}(\xi - eV)\mathcal{D}_{2\sigma}(\xi) \left[f(\xi - eV) - f(\xi)\right] d\xi, \qquad (4.5)$$

where $\mathcal{D}_{1\sigma}(\xi)$ and $\mathcal{D}_{2\sigma}(\xi)$ are the *tunneling* densities of states with spin $\sigma$ in the left and right electrodes, respectively, and $\langle|\hat{T}_{\mathbf{pk}}^{\sigma}|^2\rangle$ is the averaged tunneling probability and taken to be a constant proportional to $\exp(-2\kappa d)$, where $\kappa = \sqrt{2m\Phi}/\hbar$ is the decay constant of the wave function penetrated in the barrier and $\Phi$ is the barrier height. The total current $I$ is given by the sum of the currents in the up and down-spin channels: $I = I_\uparrow + I_\downarrow$.

Electron tunneling from ferromagnetic transition metals and alloys into a superconducting Al electrode through an insulating $Al_2O_3$ barrier is one of the most powerful tools for studying the spin-polarized electronic states of ferromagnets. In the early 1970s, Meservey, Tedrow, and coworkers [3] found that in a thin film of superconducting Al, the BCS density of states splits into up and down states by application of magnetic fields. This splitting originates from the Zeeman splitting in the quasiparticle dispersion in a magnetic field $H$:

$$E_{\mathbf{k}\sigma} = \left(\xi_{\mathbf{k}}^2 + \Delta^2\right)^{1/2} - \sigma\mu_B H, \tag{4.6}$$

where $\Delta$ is the superconducting energy gap. In the absence of spin-flip scattering in SC, the spin-dependent density of states for quasiparticles of SC in the second electrode is given by

$$\mathcal{D}_{2\uparrow}(E) = \mathcal{D}_{\text{BCS}}(E - \mu_B H), \quad \mathcal{D}_{2\downarrow}(E) = \mathcal{D}_{\text{BCS}}(E + \mu_B H), \tag{4.7}$$

where $\mathcal{D}_{\text{BCS}}(E)$ is the BCS density of states

$$\frac{\mathcal{D}_{\text{BCS}}(E)}{\mathcal{D}_N} = Re\left[\frac{|E|}{\sqrt{E^2 - \Delta^2}}\right], \tag{4.8}$$

with $\mathcal{D}_N$ the density of state of SC in the normal state. Figure 4.2(a) illustrates the splitting of the BCS density of states in a magnetic field $H$. This Zeeman splitting provides the basis for spin-polarized tunneling.

The Zeeman splitting in the density of states of SC enables one to extract the spin polarization of various ferromagnets using ferromagnet/insulator/superconductor (FM/I/SC) junctions. In magnetic fields, the densities of states of the up- and down-spin bands in FM correspond to those of the majority and minority spin bands, $\mathcal{D}_{1\uparrow} = \mathcal{D}_M$ and $\mathcal{D}_{1\downarrow} = \mathcal{D}_m$, respectively, while those in the SC electrode $\mathcal{D}_{2\sigma}$ are given by Eq. (4.7). Then, from Eq. (4.5), the conductance $G = dI/dV$ of the junction is given by the sum of the conductance for the two independent spin directions:

$$\begin{aligned} G(V)/G_N &= \left(\frac{1+P}{2}\right)\int_{-\infty}^{\infty}\frac{\mathcal{D}_{\text{BCS}}(E-\mu_B H)/\mathcal{D}_N}{\cosh^2\left[(E-eV)/2k_B T\right]}\frac{dE}{4k_B T}\\ &+ \left(\frac{1-P}{2}\right)\int_{-\infty}^{\infty}\frac{\mathcal{D}_{\text{BCS}}(E+\mu_B H)/\mathcal{D}_N}{\cosh^2\left[(E-eV)/2k_B T\right]}\frac{dE}{4k_B T}, \end{aligned} \tag{4.9}$$

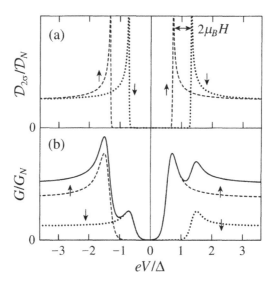

Figure 4.2: (a) Zeeman splitting of the BCS density of states into spin-up (dashed) and spin-down (dotted) densities of states in a magnetic field $H$, showing a splitting of $2\mu_B H$. (b) Spin-up conductance (dashed), spin-down conductance (dotted), and total conductance (solid curve).

where $G_N = G_N^\uparrow + G_N^\downarrow$ ($G_N^\sigma = (2\pi e^2/\hbar)\langle|\hat{T}_{\mathbf{pk}}^\sigma|^2\rangle \mathcal{D}_{1\sigma}\mathcal{D}_N$) is the conductance when SC is in the normal states and $P$ is the tunneling spin polarization defined by the relative conductance between the spin-up and spin-down channels

$$P = \frac{G_N^\uparrow - G_N^\downarrow}{G_N^\uparrow + G_N^\downarrow} = \frac{\mathcal{D}_M - \mathcal{D}_m}{\mathcal{D}_M + \mathcal{D}_m}. \tag{4.10}$$

If $\langle|\hat{T}_{\mathbf{pk}}^\sigma|^2\rangle$ is spin-independent, then $P$ is expressed in terms of the densities of states of FM as in the third term of Eq. (4.10). In Fig. 4.2(b), the conductance $G_\sigma$ for each spin direction (dashed or dotted) and the total conductance $G = G_\uparrow + G_\downarrow$ (solid curve) are shown for $P = 0.5$ and $T/T_c = 0.15$. The most striking feature of $G$ is its asymmetry around $V = 0$. The degree of the asymmetry is directly related to the value of $P$ through the weighted factors $\frac{1}{2}(1 + P)$ and $\frac{1}{2}(1 - P)$ in Eq. (4.9).

In the experiments of Tedrow and Meservey, electron tunneling between Al and ferromagnetic metals and alloys in high magnetic fields are used to measure the tunnel conductance. An analysis of the measured conductance based on Eq. (4.9) yields the spin-polarization $P$ for various ferromagnets. In Table 4.1, the spin polarization $P$ is listed for various ferromagnetic materials [11, 12, 13] recently measured by using improved junction preparation conditions including samples grown by molecular beam epitaxy (MBE).

Table 4.1: Spin polarization $P$ for various ferromagnetic metals and alloys. (a) Moodera and Mathon [11], (b) Monsma and Parkin [12], and (c) Worledge and Geballe [13]

| Materials | Ni | Co | Fe | $Ni_{80}Fe_{20}$ | $Co_{50}Fe_{50}$ | $La_{0.7}Sr_{0.3}MnO_3$ |
|---|---|---|---|---|---|---|
| $P^{(a)}$ | 33% | 45% | 44% | 48% | 51% | – |
| $P^{(b)}$ | 31% | 42% | 45% | 45% | 50% | – |
| $P^{(c)}$ | – | – | – | – | – | 72% |

As seen in Table 4.1, the values of the tunneling spin polarization $P$ are positive, i.e., $P > 0$, for all of 3$d$ ferromagnetic metals; the majority spin electrons are predominant in the tunnel current in all cases. From the tunneling experiments it has been established that the tunnel current from FM into other metals through the $Al_2O_3$ barrier are dominated by majority spins for Ni, Co, Fe, and their alloys, and the tunneling spin-polarization is correlated with the magnetic moment of the electrode. However, the positive sign of $P$ is surprising, especially for metals such as Co and Ni in which a negative polarization is expected due to the smaller density of states of the majority spin band at the Fermi level since the majority $d$ band is below the Fermi level.

Various theoretical explanations have been proposed to explain the positive value of the spin polarization. Stearns [14] tried to explain this tendency by observing that the ferromagnetic transition metal has a large fraction of free-electron-like character of $d$ electrons at the Fermi surface. Hertz and Aoi argued that the $s$ electrons are responsible for the tunneling and the tunnel currents are proportional to the density of $s$ states at the Fermi level, despite the much higher density of $d$ states in 3$d$ ferromagnetic metals [15]. Recent first-principle band calculations support that tunneling is dominated by electrons of $s$ or $p$ character; For $Al_2O_3$ barriers and Co electrodes, the positive spin polarization might be explained by the strong bonding between the $d$-orbitals of Co and the $sp$ orbitals of Al (or the $p$ orbitals of O) at the interface, which results in an almost unoccupied minority $sp$-density of states on Al (or the minority $p$-density of states on O) [16, 17].

Pioneering works to study the dependence of resistivity on the magnetic structure were carried out for magnetic tunnel junctions, which was about one decade earlier than the discovery of GMR in magnetic multilayers. In a magnetic tunnel junction, two ferromagnetic metals such as Fe, Co, and Ni are separated by a thin insulating layer. The tunnel current from one ferromagnetic metal to the other through the tunnel barrier depends on the relative orientation of the two ferromagnetic metals.

We first present a simple theoretical description for understanding the TMR observed in magnetic tunnel junctions based on the tunnel Hamiltonian method. In the low bias regime, where the bias voltage $V$ is much smaller than the band width (of order of $eV$) and the density of states is nearly constant,

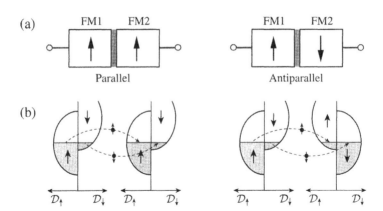

Figure 4.3: (a) Ferreomagnet/insulator/ferromagnet (FM1/I/FM2) tunnel junction in the parallel (left) and antiparallel (right) alignments of magnetizations. (b) Corresponding densities of states of FM1 and FM2 are schematically shown.

the conductance $G_\sigma = dI_\sigma/dV$ for each spin channel is given from Eq. (4.5) as

$$G_\sigma(V) \approx \frac{2\pi e^2}{\hbar} \langle |\hat{T}^\sigma_{\mathbf{pk}}|^2 \rangle \mathcal{D}_{1\sigma} \mathcal{D}_{2\sigma}, \tag{4.11}$$

where $\mathcal{D}_{1\sigma}$ and $\mathcal{D}_{2\sigma}$ are the densities of states of the spin $\sigma$ band at the Fermi levels in the ferromagnetic electrodes.

Assuming that the magnetic moments of the electrodes FM1 and FM2 are aligned antiferromagnetic (A) in zero magnetic field and aligned ferromagnetic (F) in applied magnetic fields, the total conductance $G = G_\uparrow + G_\downarrow$ in the F alignment is given by

$$G_F = G_F^\uparrow + G_F^\downarrow \propto \mathcal{D}_{M1}\mathcal{D}_{M2} + \mathcal{D}_{m1}\mathcal{D}_{m2}, \tag{4.12}$$

and in the A alignment

$$G_A = G_A^\uparrow + G_A^\downarrow \propto \mathcal{D}_{M1}\mathcal{D}_{m2} + \mathcal{D}_{m1}\mathcal{D}_{M2}, \tag{4.13}$$

where $\mathcal{D}_{Mi}$ and $\mathcal{D}_{mi}$ are the densities of states for the majority and minority spin bands in the $i$-th electrode, respectively. The TMR ratio is defined by

$$TMR = \Delta R/R_F = \frac{R_A - R_F}{R_F} = \frac{G_F - G_A}{G_A} = \frac{2P_1 P_2}{1 - P_1 P_2}, \tag{4.14}$$

with $P_i$ the spin polarization of the $i$th electrode defined by

$$P_i = \frac{\mathcal{D}_{Mi} - \mathcal{D}_{mi}}{\mathcal{D}_{Mi} + \mathcal{D}_{mi}}. \tag{4.15}$$

Note that another definition for TMR is often used, i.e., $TMR = \Delta R / R_A = 2P_1 P_2 / (1 + P_1 P_2)$.

First attempt to observe the spin-dependent effect in magnetic tunnel junction was made by Julliere in tunnel junctions Co/Ge/Ni [4] and subsequently Maekawa and Gäfvert in tunnel junctions Ni/NiO/Co, Fe and Ni [5]. They showed that the tunnel current depends on the relative orientation of the magnetic moments. These were the first realization of how a magnetic field could control the current in tunnel junctions; this concept is now called TMR. In 1975, Julliere studied a junction containing magnetic electrodes Co/Ge/Ni, and found a relatively large TMR about 14%. However, $\Delta G / G$ is attributed to an effect of a zero-bias anomaly, which is believed to be caused by magnetic impurities in the barrier. In 1978, Maekawa and Gäfvert [5] succeeded in showing the relation between the tunnel conductance and the magnetization process in a junction Ni/NiO/Co and several other ferromagnetic tunnel junctions. They have observed a TMR effect at low temperatures, however, their results did not gather intense attention until the early 1990s.

In recent years, there has been remarkable progress in fabrication technology of thin films, which makes it possible to fabricate high-quality magnetic tunnel junctions with a uniform tunnel barrier. In 1995, two groups of Miyazaki [6] and Moodera [7] demonstrated that a large TMR ratio can be achieved at room temperature. Their findings stimulate not only the basic research of the spin-dependent tunneling but also application to magnetoelectronics [18].

### 4.2.2  Free electron model

In the tunnel Hamiltonian model in the preceding section, we assumed that the tunneling matrix elements can be treated as a constant and the same for spin-up and spin-down electrons. In other words, the wavefunction in the barrier region is assumed to be independent of the wave vector and spin. However, it is not obvious that the above assumptions are justified. In 1989, Slonczewski proposed another approach for the spin dependent tunneling based on the free-electron model, where the exact wavefunction in the barrier region is used [9]. He showed that TMR is determined not only by the spin-polarization of ferromagnet electrodes, $P$, but also by the potential height of the insulating barrier. Introducing the effective spin-polarization of ferromagnetic electrode, $P_{\text{eff}}$, which ranges from $-P$ to $P$ depending on the height of barrier potential, the TMR is expressed in the same formula as that derived from the tunnel Hamiltonian model.

Before studying the spin-dependent tunneling in ferromagnetic tunnel junctions, it is convenient to consider the non-magnetic tunnel junctions and derive an approximate form of the tunnel conductance in the weak transmitting limit [8]. When the bias voltage $V$ is applied, electrons incident from the left electrode tunnel through the insulating barrier, resulting in the tunnel current as

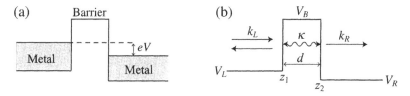

Figure 4.4: (a) Potential energy diagram for a metal/insulator/metal tunnel junction with the bias voltage $V$. The shaded area represents the occupied states of electrons at zero temperature. Electrons are transmitted from the occupied states in the left electrode to the unoccupied states in the right electrode. (b) The geometry of one-electron potential for the Hamiltonian given by Eq. (4.19). The left and right boundary of the insulating barrier is indicated by $z_1$ and $z_2$, respectively.

shown in Figs. 4.4(a) and (b). We assume that the system has translational symmetry in the transverse ($x$ and $y$) direction and therefore the wave vector parallel to the barrier surface $k_\| = (k_x, k_y)$ is conserved during the tunneling. We also assume that the temperature is zero. The number of electrons incident from the left electrode per unit time with wave vector $k_\|$ is given by

$$N(k_\|) = 2 \cdot \frac{1}{2} \int_0^{eV} dE v_z(E, k_\|) \frac{1}{\pi \hbar v_z(E, k_\|)} = 2\frac{eV}{h}, \qquad (4.16)$$

where $v_z(k_\|)$ is the velocity along the $z$ direction and $1/\pi \hbar v_z(k_\|)$ is the corresponding 1D density of states for the one spin channel. The factor 2 in Eq. (4.16) is due to the spin-degeneracy of the energy bands in the non-magnetic electrode. Note that the number of incident electrons is independent of the wave vector $k_\|$. The tunnel current density is obtained by summing up the number of electrons tunneling through the barrier:

$$I = e \int \frac{dk_\|^2}{(2\pi)^2} N(k_\|) T(k_\|) = 2\frac{e^2 V}{h} \int \frac{dk_\|^2}{(2\pi)^2} T(k_\|), \qquad (4.17)$$

where $T(k_\|)$ is the transmission probability defined as the ratio between the probability current densities of incident and transmitting waves. Here we assume the bias-voltage $V$ is so small that we can neglect the energy dependence of the transmission probability. The differential conductance per unit area is written in terms of the transmission probability as,

$$G \equiv \frac{dI}{dV} = 2\frac{e^2}{h} \int \frac{dk_\|^2}{(2\pi)^2} T(k_\|). \qquad (4.18)$$

The transmission probability $T(k_\|)$ is obtained by solving the 1D Schrödinger equation:

$$-\frac{\hbar^2}{2m} \frac{\partial^2}{\partial z^2} \psi(z) = \left( E - V(z) - \frac{\hbar^2}{2m} k_\|^2 \right) \psi(z), \qquad (4.19)$$

in the geometry shown in Fig. 4.4(b). The general solutions of Eq. (4.19) in the left $(L)$ electrode, barrier $(B)$, and right $(R)$ electrode are, respectively, of the forms

$$\psi_L(z) = a_L e^{ik_L z} + b_L e^{-ik_L z}, \qquad \text{for } z \leq z_1,$$

$$\psi_B(z) = a_B e^{\kappa z} + b_B e^{-\kappa z}, \qquad \text{for } z_1 < z \leq z_2, \qquad (4.20)$$

$$\psi_R(z) = a_R e^{ik_R z} + b_R e^{-ik_R z}, \qquad \text{for } z > z_2,$$

where the $z$ component of the wave numbers are defined as

$$k_L = \sqrt{(2m/\hbar^2)(E - V_L) - k_\parallel^2},$$

$$\kappa = \sqrt{k_\parallel^2 - (2m/\hbar^2)(E - V_B)}, \qquad (4.21)$$

$$k_R = \sqrt{(2m/\hbar^2)(E - V_R) - k_\parallel^2}.$$

The scattering wave in the whole system is given by the combination of eigen functions in Eq. (4.20). The coefficients $a_L$, $b_L$, $a_B$, $b_B$, $a_R$, and $b_R$ are determined by matching the slope and value of the wavefunction across the interface [8]. The matching conditions at $z = z_1$ are conveniently described as a $2 \times 2$ matrix $R_1$ operating on the 2D vectors as

$$\begin{pmatrix} a_L \\ b_L \end{pmatrix} = R_1 \begin{pmatrix} a_B \\ b_B \end{pmatrix}, \qquad (4.22)$$

where

$$R_1 = \frac{1}{2k_L} \begin{pmatrix} (k_L - i\kappa)e^{(-ik_L + \kappa)z_1} & (k_L + i\kappa)e^{(-ik_L - \kappa)z_1} \\ (k_L + i\kappa)e^{(ik_L + \kappa)z_1} & (k_L - i\kappa)e^{(ik_L - \kappa)z_1} \end{pmatrix}. \qquad (4.23)$$

In the same manner, the matching conditions at $z = z_2$ are written as

$$\begin{pmatrix} a_B \\ b_B \end{pmatrix} = R_2 \begin{pmatrix} a_R \\ b_R \end{pmatrix}, \qquad (4.24)$$

where

$$R_2 = \frac{i}{2\kappa} \begin{pmatrix} (k_R - i\kappa)e^{(ik_R - \kappa)z_2} & -(k_R + i\kappa)e^{-(ik_R + \kappa)z_2} \\ -(k_R + i\kappa)e^{(ik_R + \kappa)z_2} & (k_R - i\kappa)e^{-(ik_R - \kappa)z_2} \end{pmatrix}. \qquad (4.25)$$

Since the quantity that we wish to compute is the transmission probability for the electron incident from the left electrode, only a transmitted wave exists in the right electrode and $b_R = 0$. The relation between the coefficients $a_L$, $b_L$, and $a_R$ is

$$\begin{pmatrix} a_L \\ b_L \end{pmatrix} = R_1 R_2 \begin{pmatrix} a_R \\ 0 \end{pmatrix}. \qquad (4.26)$$

Thus, the transmission probability is

$$T(\boldsymbol{k}_{\|}) = \frac{|a_R|^2 k_R}{|a_L|^2 k_L} = \frac{k_R}{k_L} \frac{1}{|(\mathsf{R}_1\mathsf{R}_2)_{11}|^2} \simeq \frac{16 k_L \kappa^2 k_R}{(k_L^2 + \kappa^2)(k_R^2 + \kappa^2)} e^{-2\kappa d}, \quad (4.27)$$

where $d = z_2 - z_1$ is the thickness of the barrier. In the last equality of Eq. (4.27), we use the condition $e^{-2\kappa d} \ll 1$ because we are interested in the weak transmitting limit $T(\boldsymbol{k}_{\|}) \ll 1$. Substituting Eq. (4.27) into Eq. (4.18), the conductance is written as

$$G = 2\frac{e^2}{h} \int \frac{dk_{\|}^2}{(2\pi)^2} \frac{16 k_L \kappa^2 k_R}{(k_L^2 + \kappa^2)(k_R^2 + \kappa^2)} e^{-2\kappa d}. \quad (4.28)$$

We introduce the symbols $k_{L0}$, $k_{R0}$ and $\kappa_0$ to represent the wave numbers with $\boldsymbol{k}_{\|} = 0$. In Eq. (4.28), the value of $k_{\|}$ is limited in the range of $0 \le k_{\|} < \min[k_{L0}, k_{R0}]$. For the high barrier, $\kappa_0 \gg k_{\|}$, and the wave number $\kappa$ can be expressed as

$$\kappa = \kappa_0 \left[ 1 + \frac{1}{2} \left( \frac{k_{\|}}{\kappa_0} \right)^2 \right] = \kappa_0 + \frac{k_{\|}^2}{2\kappa_0}. \quad (4.29)$$

Since we consider an elliptical energy band, it is often convenient to use

$$\int \frac{d^2 k_{\|}}{(2\pi)^2} = \int \rho_{\|}(E_{\|}) dE_{\|}, \quad (4.30)$$

where $E_{\|} = (\hbar^2/2m)k_{\|}^2$ and the 2D density of states $\rho_{\|}(E_{\|}) = (m/2\pi\hbar^2)$. Therefore, the conductance can be written as

$$G = 2\frac{e^2}{h}\rho_{\|} \int_0^{\frac{\hbar^2}{2m}\min[k_{L0}^2, k_{R0}^2]} T(E_{\|}) dE_{\|}, \quad (4.31)$$

where

$$\begin{aligned} T(E_{\|}) &= \frac{16 e^{-2\kappa_0 d}}{(V_L + V_B)(V_R + V_B)} \sqrt{\frac{\hbar^2}{2m} k_{L0}^2 - E_{\|}} \sqrt{\frac{\hbar^2}{2m} k_{R0}^2 - E_{\|}} \\ &\quad \times \left( \frac{\hbar^2}{2m} \kappa_0^2 + E_{\|} \right) \exp\left( -\frac{d}{\kappa_0} \frac{2m}{\hbar^2} E_{\|} \right). \end{aligned} \quad (4.32)$$

For large $d$, the important region in the integral of Eq. (4.31) is $E_{\|} \simeq 0$ and we have

$$G = 2\frac{e^2}{h}\frac{\kappa_0}{4\pi d} T(0) = 2\frac{e^2}{h}\frac{4\kappa_0}{\pi d}\frac{k_{L0}\kappa_0^2 k_{R0}}{(k_{L0}^2 + \kappa_0^2)(k_{R0}^2 + \kappa_0^2)} e^{-2\kappa_0 d}. \quad (4.33)$$

The conductance for spin-up and spin-down electrons is given by the half of the total conductance in Eq. (4.33) as,

$$G_{\uparrow} = G_{\downarrow} = \frac{e^2}{h}\frac{4\kappa_0}{\pi d}\frac{k_{L0}\kappa_0^2 k_{R0}}{(k_{L0}^2 + \kappa_0^2)(k_{R0}^2 + \kappa_0^2)} e^{-2\kappa_0 d}. \quad (4.34)$$

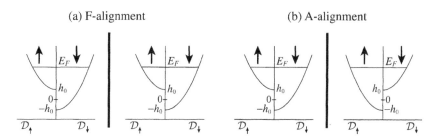

Figure 4.5: Densities of states for spin-up ($\mathcal{D}_\uparrow$) and spin-down ($\mathcal{D}_\downarrow$) electrons in the F alignment (a) and those in the A alignment (b).

Let us move onto the magnetic tunnel junctions, where the conductance is spin-dependent. In the magnetic electrodes, spin-up and spin-down electrons feel the different exchange potential $h_0$ and $-h_0$, respectively. The matching condition depends on the relative angle between the magnetization vectors of left and right electrodes. For simplicity, we consider the two types of alignments of magnetization vectors, ferromagnetic (F) and antiferromagnetic (A) alignments as shown in Figs. 4.5(a) and (b). The spin quantization axis is taken to be parallel to the magnetization vector in the left electrode. For the F alignment, where the magnetization vectors are parallel, we have two tunneling processes: electrons incident from the left majority (minority) spin band tunnel to the right majority (minority) spin band. The potential diagram for both tunneling processes are shown in Figs. 4.6(a) and (b), respectively. The wave number of electrons in the majority and minority spin bands are

$$k_M = \sqrt{(2m/\hbar^2)(E - h_0) - k_\parallel^2} \quad \text{(majority)} \qquad (4.35)$$

$$k_m = \sqrt{(2m/\hbar^2)(E + h_0) - k_\parallel^2} \quad \text{(minority)}. \qquad (4.36)$$

The conductance for spin-up (spin-down) electrons is obtained from Eq. (4.34) by setting $k_{L0} = k_{R0} = k_{M0}(k_{m0})$, where $k_{M0}(k_{m0})$ is the wave number for the electrons in the majority (minority) spin-band with $k_\parallel = 0$:

$$G_\uparrow^F = \frac{e^2}{h} \frac{4\kappa_0^3}{\pi d} \frac{k_{m0}^2}{(k_{m0}^2 + \kappa_0^2)^2} e^{-2\kappa_0 d}, \quad G_\downarrow^F = \frac{e^2}{h} \frac{4\kappa_0^3}{\pi d} \frac{k_{M0}^2}{(k_{M0}^2 + \kappa_0^2)^2} e^{-2\kappa_0 d}. \qquad (4.37)$$

The total conductance for the F alignment takes the form

$$G^F = G_\uparrow^F + G_\downarrow^F = \frac{e^2}{h} \frac{4\kappa_0^3}{\pi d} \left[ \frac{k_{m0}^2}{(k_{m0}^2 + \kappa_0^2)^2} + \frac{k_{M0}^2}{(k_{M0}^2 + \kappa_0^2)^2} \right] e^{-2\kappa_0 d}. \qquad (4.38)$$

On the contrary, for the A alignment where the magnetization vectors are antiparallel, electrons incident from the left majority (minority) spin-band

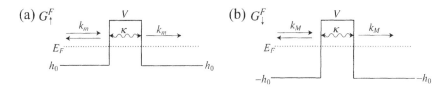

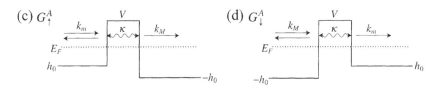

Figure 4.6: Geometries of the potentials for spin-up and spin-down electrons in the F alignment are shown in panels (a) and (b), and those in the A alignment are in panels (c) and (d), respectively.

tunnel to the right minority (majority) spin-band. The potential diagrams for spin-up and spin-down electrons are depicted in Figs. 4.6(c) and 4.6(d), respectively. The conductance for the A alignment is given by

$$G^A_\uparrow = G^A_\downarrow = \frac{e^2}{h} \frac{4\kappa_0^3}{\pi d} \frac{k_{m0}k_{M0}}{(k_{m0}^2 + \kappa_0^2)(k_{M0}^2 + \kappa_0^2)} e^{-2\kappa_0 d}, \qquad (4.39)$$

$$G^A = G^A_\uparrow + G^A_\downarrow = \frac{e^2}{h} \frac{4\kappa_0^3}{\pi d} \frac{2k_{m0}k_{M0}}{(k_{m0}^2 + \kappa_0^2)(k_{M0}^2 + \kappa_0^2)} e^{-2\kappa_0 d}. \qquad (4.40)$$

Using Eqs. (4.38) and (4.40), the TMR is expressed as [9],

$$\begin{aligned} TMR &= \frac{G^F - G^A}{G^A} \\ &= \frac{(k_{m0} - k_{M0})^2(\kappa_0^2 - k_{m0}k_{M0})^2}{2k_{m0}k_{M0}(k_{m0}^2 + \kappa_0^2)(k_{M0}^2 + \kappa_0^2)} \\ &= \frac{2P_{\text{eff}}^2}{1 - P_{\text{eff}}^2}, \end{aligned} \qquad (4.41)$$

where $P_{\text{eff}}$ is the effective spin-polarization defined as

$$P_{\text{eff}} = \frac{\kappa_0^2 - k_{m0}k_{M0}}{\kappa_0^2 + k_{m0}k_{M0}} P, \qquad (4.42)$$

in which $P = (k_{M0} - k_{m0})/(k_{M0} + k_{m0})$ is the spin-polarization of the ferromagnetic electrode with an elliptical energy band.

Since the wave number in the barrier region $\kappa_0$ ranges from 0 (low barrier limit) to $\infty$ (high barrier limit), we have $-P < P_{\text{eff}} < P$. In the high

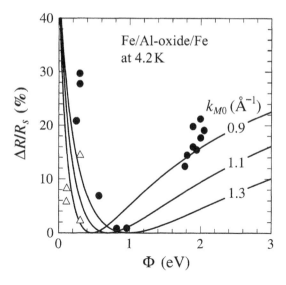

Figure 4.7: Dependence of TMR on the barrier height $\Phi$ at 4.2 K [19]. The horizontal axis is the barrier height obtained by fitting the $I$-$V$ curves to the Simmon's relation [20]. Solid curves are the calculated values using the free electron model.

barrier limit, the TMR is the same as that obtained by the tunneling Hamiltonian method. One interesting result of the free electron model is to diminish $P_{eff}$ and therefore the TMR for values of the barrier height ($\kappa_0 \sim k_{m0}k_{M0}$). This variation of the TMR has been experimentally observed by Tezuka and Miyazaki as shown in Fig. 4.7 [19].

## 4.2.3 Tight binding model

Recently the more realistic calculation based on tight-binding Hamiltonian and the Kubo-Landauer formula have been developed by several authors [10,11,21-27]. The tight-binding model is of great use both in dealing with the multi-orbital systems with the realistic band structure and in studying the effect of the disorder and roughness of the insulating barrier on the spin-dependent transport. In order to understand the basic formalism of the conductance calculation in the tight binding model, we first consider a 1D chain shown in Fig. 4.8, which is expressed by the following 1D Hamiltonian:

$$\mathcal{H} = -t \sum_{\langle n,n' \rangle} c_n^\dagger c_{n'} + \sum_n \varepsilon_n c_n^\dagger c_n, \qquad (4.43)$$

where $t$ is the hopping matrix element and $\varepsilon_n$ is the on-site energy at $n$th site. The spin indices are dropped for the simplicity. In the low bias regime, the

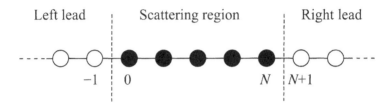

Figure 4.8: Schematic of the 1D chain including the scattering region, which is indicated by filled circles. The scattering region is connected with the left and right ideal lead.

conductance $G$ at zero temperature is given by the Kubo formula as

$$G = \frac{\pi\hbar}{N^2}\sum_{\alpha\beta}\left|\sum_{n=1}^{N}\langle\alpha|J(n)|\beta\rangle\right|^2 \delta(E_F - E_\alpha)\delta(E_F - E_\beta), \qquad (4.44)$$

where $|\alpha\rangle$ and $|\beta\rangle$ are the eigenstates of the system with energies $E_\alpha$ and $E_\beta$, respectively, and the current operator at site $n$ is defined as

$$J(n) = \frac{ie}{\hbar}\left[tc_{n+1}^\dagger c_n - tc_n^\dagger c_{n+1}\right]. \qquad (4.45)$$

In the stationary state of the system, the current conservation requires that $J(n)$ is independent of $n$, and thus the conductance reduces to

$$\begin{aligned}
G &= \frac{\pi e^2}{\hbar}t^2\sum_{\alpha\beta}\Big[\langle n+1|\alpha\rangle\langle\alpha|n+1\rangle\langle n|\beta\rangle\langle\beta|n\rangle \\
&\quad - \langle n|\alpha\rangle\langle\alpha|n+1\rangle\langle n|\beta\rangle\langle\beta|n+1\rangle \\
&\quad - \langle n+1|\alpha\rangle\langle\alpha|n\rangle\langle n+1|\beta\rangle\langle\beta|n\rangle \\
&\quad + \langle n|\alpha\rangle\langle\alpha|n\rangle\langle n+1|\beta\rangle\langle\beta|n+1\rangle\Big] \\
&\quad \times \delta(E_F - E_\alpha)\delta(E_F - E_\beta), \qquad (4.46)
\end{aligned}$$

where $|n\rangle$ is the one-particle state created by the operator $c_n^\dagger$. Let us introduce the Green's function:

$$\tilde{\mathcal{G}}(n,n') = \frac{i}{2}\left[\mathcal{G}^+(n,n') - \mathcal{G}^-(n,n')\right] = \pi\sum_\alpha\langle n|\alpha\rangle\langle\alpha|n'\rangle\delta(E - E_\alpha), \quad (4.47)$$

where $\mathcal{G}^+$ and $\mathcal{G}^-$ are the retarded and advanced one-electron Green's functions expressed as

$$\mathcal{G}^\pm(n,n') = \langle n|\frac{1}{E - \mathcal{H} \pm i\delta}|n'\rangle. \qquad (4.48)$$

Here, $\mathcal{H}$ is the total Hamiltonian and $\delta$ is the small positive number. Using the Green's function defined by Eq. (4.47), the conductance can be written as

$$G = \frac{4e^2}{h}\left[t\tilde{\mathcal{G}}(n,n)t\tilde{\mathcal{G}}(n+1,n+1) - t\tilde{\mathcal{G}}(n,n+1)t\tilde{\mathcal{G}}(n,n+1)\right]. \qquad (4.49)$$

The advanced and retarded Green's functions are easily obtained by the recursion method [28, 29]. In the matrix form, the Hamiltonian is given by

$$
H = \begin{pmatrix}
\ldots & -t & 0 & 0 & 0 \\
-t & \varepsilon_0 & -t & 0 & 0 \\
0 & -t & \varepsilon_1 & -t & 0 \\
0 & 0 & -t & \varepsilon_2 & -t \\
0 & 0 & 0 & -t & \ldots
\end{pmatrix},
\tag{4.50}
$$

where the dimension of the matrix is infinite since we consider the open system. Let us define the Green's function for the system in which all sites $n > n_0$ is deleted as,

$$
G_{n_0}^{L\pm} = \left[ (E \pm i\eta)\mathsf{I} - H_{n_0}^L \right]^{-1},
\tag{4.51}
$$

where the Hamiltonian $H_{n_0}^L$ consists of sites $n \geq n_0$. The Green's function for the system $H_{n_0+1}^L$ satisfies the following equation:

$$
\left[ (E \pm i\eta)\mathsf{I} - H_{n_0+1}^L \right] G_{n_0+1}^{L\pm} = \mathsf{I},
\tag{4.52}
$$

$$
\begin{pmatrix}
(E \pm i\eta)\mathsf{I} - H^L(n_0) & -t \\
-t^\dagger & E \pm i\eta - \varepsilon_{n_0+1}
\end{pmatrix}
\begin{pmatrix}
A_1 & A_2 \\
A_3 & \mathcal{G}_{n_0+1}^{L\pm}(n_0+1, n_0+1)
\end{pmatrix} = \mathsf{I},
\tag{4.53}
$$

where $t^\dagger = (0, \cdots, 0, t)$. It follows that the Green's function $\mathcal{G}_n^{L\pm}(n, n)$ satisfies the following recursive relation:

$$
\mathcal{G}_{n_0+1}^{L\pm}(n_0+1, n_0+1) = \left[ g^\pm(n_0+1)^{-1} - t\mathcal{G}_{n_0}^{L\pm}(n_0, n_0)t \right]^{-1},
\tag{4.54}
$$

where $g^\pm(n) = (E \pm i\eta - \varepsilon_{n_0+1})^{-1}$ is the Green's function for the isolated site $n_0 + 1$. Starting from the left semi-infinite ideal lead, where the exact Green's function is analytically given, we can evaluate the Green's function $\mathcal{G}_n^{L\pm}(n, n)$ at arbitrary site $n \geq 1$.

The Green's function for the whole system satisfies

$$
\begin{pmatrix}
(E \pm i\eta)\mathsf{I} - H_{n-1}^L & -t & 0 \\
-t^\dagger & E \pm i\eta - \varepsilon_n & -t \\
0 & -t^\dagger & (E \pm i\eta)\mathsf{I} - H_{n+1}^R
\end{pmatrix}
$$
$$
\times \begin{pmatrix}
A_1 & A_2 & A_3 \\
A_4 & \mathcal{G}^\pm(n, n) & A_5 \\
A_6 & A_7 & A_8
\end{pmatrix} = \mathsf{I},
\tag{4.55}
$$

where $H_{n_0}^R$ is the Hamiltonian for the semi-infinite system in which all sites $n < n_0$ is deleted. The Green's function for the isolated right electrode is calculated by using the following recursive relation:

$$
\mathcal{G}_{n_0}^{R\pm}(n_0, n_0) = \left[ g^\pm(n_0)^{-1} - t\mathcal{G}_{n_0+1}^{R\pm}(n_0+1, n_0+1)t \right]^{-1}.
\tag{4.56}
$$

The diagonal element of the advanced and retarded Green's function for the whole system $\mathcal{G}^{\pm}(n,n)$ is given by

$$\mathcal{G}^{\pm}(n,n) = \left[g^{\pm}(n)^{-1} - t\mathcal{G}_{n-1}^{L\pm}(n-1,n-1)t - t\mathcal{G}_{n+1}^{R\pm}(n+1,n+1)t\right]^{-1}.$$
(4.57)

It is also easy to show that the off-diagonal element is given by

$$\mathcal{G}^{\pm}(n,n+1) = \mathcal{G}^{\pm}(n,n)t\mathcal{G}\pm_{n+1}^{R}(n+1,n+1).$$
(4.58)

Finally, the total conductance for each spin-direction is calculated by using Eqs. (4.49)-(4.58).

The tight-binding method described above is easily generalized to the 3D junction with multi-orbital system. The Hamiltonian is given by

$$\mathcal{H} = -\sum_{\langle n,n'\rangle,\alpha,\beta,\sigma} t_{n\alpha,n'\beta}c_{n\alpha\sigma}^{\dagger}c_{n'\beta\sigma} + \sum_{n,\alpha,\sigma}\varepsilon_{n\alpha\sigma}c_{n\alpha\sigma}^{\dagger}c_{n\alpha\sigma},$$
(4.59)

where the lattice sites are labeled by indices $n$ and $n'$, and $c_{n\alpha\sigma}^{\dagger}(c_{n\alpha\sigma})$ is the creation (annihilation) operator of an electron with spin $\sigma$ and orbital $\alpha$ on the lattice site $n$. The hopping energy is $t_{nn'\alpha\beta}$ and the summation $\langle n,n'\rangle$ runs over nearest-neighbor sites. The spin-dependent on-site potential $\varepsilon_{n\alpha\sigma}$ includes the exchange potential in the ferromagnet and the disorder and roughness of the insulating barrier are described by the variation of the hopping $t_{n\alpha,n'\beta}$ and on-site energy $\varepsilon_{n\alpha\sigma}$ [25].

Mathon has studied the continuous transition from the CPP GMR of metallic systems to the TMR of tunnel junctions with a vacuum gap and with an insulating barrier [22]. The tunnel junction with a vacuum gap is obtained by gradually turning off the overlap matrix elements between the ferromagnetic electrodes. The insulating barrier is obtained by varying the on-site potentials in the spacer so that the Fermi level in the spacer layer moves into the band gap. It is shown that the tunneling across the vacuum gap and through an insulating barrier leads to the same TMR provided the barrier is at least as high as the conduction band width and the barrier is narrow, not wider than a few atomic planes. It is also shown that the tunneling current across the vacuum gap is carried only by the *s-p* electrons in the tunneling regime though a significant proportion of the current in the magnetic metal (Co) is carried by *d* electrons that are highly spin polarized. The switching from *d* electrons to *s-p* electrons can explain the value of the TMR and the change in the sign of spin-polarization $P$ of tunneling electrons.

Itoh *et al.* have studied the TMR for the half-metallic systems by using the double exchange model [27]. It is shown that the strong exchange coupling in the double exchange model plays an important role in the temperature dependence of both $P$ and TMR; their values can be less than the maximum values expected for half-metallic systems at low temperatures, and the TMR

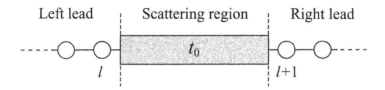

Figure 4.9: Schematic of the one-dimensional chain where the scattering region is represented by the single hopping matrix element $t_0$.

decreases more rapidly than $P$ with increasing temperature. The calculated results, however, indicate that the TMR ratio may still be large at high temperatures near the Curie temperature.

The effect of disorder and roughness of the insulating barrier on TMR has been studied by several authors [10, 22-25]. It is obvious that the free electron model can be expressed as the single orbital tight-binding model with specular barrier ($\boldsymbol{k}_\parallel$ is conserved) by discretizing the Schrödinger equation. On the other hand, when the coherence across the barrier is completely lost due to the randomness and roughness of the insulating barrier and the electron hopping between the electrodes is weak, the conductance and therefore MR in the tight-binding model reduce to the usual expressions for those obtained in the tunnel Hamiltonian model.

Let us consider a 1D chain where the left and right ideal leads are connected by the scattering region represented by the single hopping matrix element $t_0$ as shown in Fig. 4.9. The conductance in the tight binding model can be written as

$$G = \frac{4e^2}{h} \left[ t_0^2 \tilde{\mathcal{G}}(\ell, \ell) \tilde{\mathcal{G}}(\ell+1, \ell+1) - t_0^2 \tilde{\mathcal{G}}(\ell, \ell+1) \tilde{\mathcal{G}}(\ell, \ell+1) \right]. \qquad (4.60)$$

By using the surface Green's function $\mathcal{G}_\ell^{L(R)\pm}(\ell, \ell)$ for the left (right) ideal lead, the Green's functions $\tilde{\mathcal{G}}(\ell, \ell)$, $\tilde{\mathcal{G}}(\ell+1, \ell+1)$, $\tilde{\mathcal{G}}(\ell, \ell+1)$ and $\tilde{\mathcal{G}}(\ell+1, \ell)$ become

$$\mathcal{G}^\pm(\ell, \ell) = \frac{\mathcal{G}_\ell^{L\pm}(\ell, \ell)}{1 - t_0^2 \mathcal{G}_\ell^{L\pm}(\ell, \ell) \mathcal{G}_{\ell+1}^{R\pm}(\ell+1, \ell+1)}, \qquad (4.61)$$

$$\mathcal{G}^\pm(\ell+1, \ell+1) = \frac{\mathcal{G}_{\ell+1}^{R\pm}(\ell+1, \ell+1)}{1 - t_0^2 \mathcal{G}_\ell^{L\pm}(\ell, \ell) \mathcal{G}_{\ell+1}^{R\pm}(\ell+1, \ell+1)}, \qquad (4.62)$$

$$\mathcal{G}^\pm(\ell, \ell+1) = \mathcal{G}^\pm(\ell+1, \ell) = \frac{t_0 \mathcal{G}_\ell^{L\pm}(\ell, \ell) \mathcal{G}_{\ell+1}^{R\pm}(\ell+1, \ell+1)}{1 - t_0^2 \mathcal{G}_\ell^{L\pm}(\ell, \ell) \mathcal{G}_{\ell+1}^{R\pm}(\ell+1, \ell+1)}. \qquad (4.63)$$

After some calculations, the conductance is written as

$$\begin{aligned} G = {} & \frac{4e^2 \pi^2}{h} \frac{t_0^2}{|1 - t_0^2 \mathcal{G}_\ell^{L-}(\ell, \ell) \mathcal{G}_{\ell+1}^{R-}(\ell+1, \ell+1)|^2} \\ & \times \left[ \frac{1}{\pi} \operatorname{Im} \mathcal{G}_\ell^{L-}(\ell, \ell) \right] \left[ \frac{1}{\pi} \operatorname{Im} \mathcal{G}_{\ell+1}^{R-}(\ell+1, \ell+1) \right]. \end{aligned} \qquad (4.64)$$

When the electron hopping between the electrodes is weak $t_0 \simeq 0$, the term

$$\frac{t_0^2}{|1 - t_0^2 \mathcal{G}_\ell^{L-}(\ell, \ell) \mathcal{G}_{\ell+1}^{R-}(\ell + 1, \ell + 1)|^2} \tag{4.65}$$

reduces to be $t_0^2$. Moreover, the densities of states in the isolated left and right electrodes are $\mathcal{D}_L = \frac{1}{\pi} \mathrm{Im} \mathcal{G}_\ell^{L-}(\ell, \ell)$ and $\mathcal{D}_R = \frac{1}{\pi} \mathrm{Im} \mathcal{G}_{\ell+1}^{R-}(\ell+1, \ell+1)$, respectively. Finally, the conductance is expressed in the same form as that obtained by the tunneling Hamiltonian model (Eq. (4.49))

$$G = \frac{2\pi e^2}{\hbar} t_0^2 \mathcal{D}_L \mathcal{D}_R. \tag{4.66}$$

For 3D systems, the following approximations are involved [24]. (i) Given that $t \simeq 0$ (electron hopping between the electrodes is weak), it is assumed that the Kubo formula can be linearized, (ii) Only tunneling between the same orbital is considered and assumed to be equally probable, i.e., the hopping matrix $t$ is replaced by $t_0 \mathsf{I}$, where $\mathsf{I}$ is a unit matrix in the orbital space and $t_0$ is a single tunneling matrix element independent of $k_\parallel$. (iii) Complete loss of the coherence across the barrier is imposed, i.e., it is assumed that a state $k_\parallel$ tunnels with an equal probability to any other state $k_\parallel'$. With these approximations, the conductance takes the form

$$\begin{aligned}
G &= \frac{2\pi e^2}{\hbar N_\parallel} \left[ \sum_{\boldsymbol{k}_\parallel} \mathrm{Tr} \; \mathrm{Im} \mathcal{G}_\ell^{L-}(\ell, \ell; \boldsymbol{k}_\parallel) \right] \left[ \sum_{\boldsymbol{k}_\parallel} \mathrm{Tr} \; \mathrm{Im} \mathcal{G}_\ell^{R-}(\ell, \ell; \boldsymbol{k}_\parallel) \right] \\
&= \frac{2\pi e^2}{\hbar N_\parallel} t_0^2 \mathcal{D}_L \mathcal{D}_R, \tag{4.67}
\end{aligned}$$

where the trace is over the orbital indices and $N_\parallel$ is the number of atoms in the plane of the junction.

### 4.2.4   Some new directions

Recently, Fert *et al.* have demonstrated that the choice of the barrier influences strongly and even reverse the spin-polarization of tunneling electrons from a given ferromagnetic metal [30]. In contrast to the results for $Al_2O_3$ barriers, in which the effective spin polarization of Co is positive at the interface between a Co electrode and a $Al_2O_3$ barrier, the effective spin polarization of Co is negative at the interface between a Co electrode and a $SrTiO_3$ barrier. These results indicate that the bonding effect at the electrode/barrier interface controls the effective spin polarization of a given electrode and may explain the influence of the choice of barriers.

It is of crucial importance for a large and reproducible TMR to make magnetic tunnel junctions with a fully oxidized and flat tunnel barrier. Yuasa *et al.* [31] have made an atomically flat magnetic tunnel junctions using an MBE technique, and studied $Fe/Al_2O_3/Fe_{50}Co_{50}$ tunnel junctions with the electrodes of single crystal Fe oriented to the (100), (110) and (211) surfaces. They found that the averaged MR ratios of these junctions at $T = 2\,K$ are 13% for Fe (100), 32% for Fe (110), and 42% for Fe (211). Such a strong dependence of TMR on the crystal orientation originates from the anisotropy of the Fe spin polarization in the momentum space and the momentum selective tunneling dominated by electrons with wave vector normal to the barrier (momentum filtering effect).

Further information on recent developments can be found in review articles [11, 32].

## 4.3  Effects of Coulomb blockade on TMR

Recent advances in nanofabrication techniques have led to a number of experiments which exhibit the Coulomb blockade (CB) phenomenon in nanostructure materials and devices: granular films, lithographically patterned tunnel junctions, and small metal droplets probed by STM. The dimension of these systems is very small (of the order of a few nanometers or submicrometers) so that the charge of a single electron plays a significant role in electron transport. To illustrate the role of charging effect in transport, we consider a small granule between two electrodes as shown in Fig. 4.10. The granule is often called *island* because the grain is surrounded by an insulating material. Let an electron jump onto a small island from the left electrode. For a very small island, a single electron tunneling charges the island by $e$ and increases the electrostatic charging energy by $e^2/2C$, where $e$ is the electronic charge and $C$ is the capacitance of the island. Therefore, unless the charging energy is overcome by bias potential $(eV)$ or thermal energy $(k_B T)$, an electron is not able to propagate between the electrodes. This is called the CB.

In this section, we study the interplay between spin-dependent tunneling and the CB in magnetic nanostructures. We first explain the basic concept of

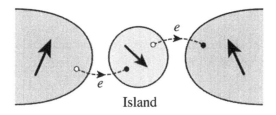

Figure 4.10: Electron tunneling between electrodes through an *island* is blocked if the charging energy of a single excess electron on the island is large in comparison with the energy of bias potential and/or thermal fluctuations.

the CB in a system with a metal-droplet probed by STM, based on the classical theory of CB. We show that a step-like structure called Coulomb staircase appears in the *I-V* curve in a magnetic granular films measured by STM, and provides important informations on individual granules. Then, we proceed to discuss the effect of higher-order tunneling (cotunneling) in single-electron transistors and granular films, where several tunneling events across different junctions or between granules occurs coherently, and show that cotunneling gives rise to a large enhancement of TMR.

### 4.3.1   Basic concept of Coulomb blockade

Let us consider the charge transport through a small metal island (droplet) between an STM-tip and a base electrode as shown in Fig. 4.11(a). The tunnel junctions are formed between the STM tip and a small metal droplet and between the droplet and the base electrode. The junction parameters such as the tunnel resistance and the capacitance are varied by changing the position of the STM tip on the island. The single electron charging effects are observed by positioning the STM above the small island.

To derive the basic equations for calculating the STM current, the system is modeled as a generic double tunnel junction shown in Fig. 4.11(b). Each junction is characterized by the tunnel resistance $R_i$ and the junction capacitance $C_i$, and carries the charge $Q_i$ $(i = 1, 2)$. The island charge is $Q_2 - Q_1$ and is changed only by tunneling of electrons into or out of the island, leading to the quantization of the island charge

$$n = (Q_1 - Q_2)/e. \tag{4.68}$$

Note that the sign of $e$ is taken to be positive (the charge of the electron is $-e$). The difference in the electrostatic energies of the charge states, $n$ and $n \pm 1$, before and after the tunneling process plays a crucial role in the CB phenomena of the system. For the moment, we ignore the spin degree of freedom to elucidate the effect of CB on the charge transport. The spin-dependent transport in the metal-droplet system is discussed in Sec. 4.3.1.2.

To evaluate the tunneling rate, we first consider the change in the electrostatic energy $E_i^{\pm}(n)$ for the transition of the island charge state $n \to n \pm 1$ by tunneling of an electron through the $i$-th junction. The free energy of the charge state characterized by $n$ is given by

$$F(n, \xi) = \frac{Q_1^2}{2C_1} + \frac{Q_2^2}{2C_2} - (Q_1 - e\xi)V, \tag{4.69}$$

where $\xi$ is the number of electrons supplied by the voltage source [34]. When an electron tunnels into the island through the 1st junction, for instance, the island charge changes from $-ne$ to $-(n+1)e$ and the charges on the capacitors

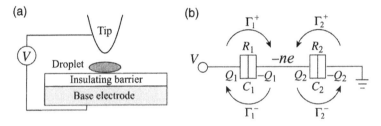

Figure 4.11: (a) Schematic figure of a tunnel junction consisting of an STM-tip, a metal droplet and a base electrode [33]. The Al oxide layer is used as a tunnel barrier between an In droplet and Al base electrode in Ref. [33]. Electrons tunnel from the STM-tip to the base electrode through the droplet. (b) Equivalent circuit consisting of two tunnel junctions. Each junction is characterized by the tunnel resistance $R_i$ and the capacitance $C_i$, carrying the charge $Q_i$. The island is charged by $-ne$.

deviate from their initial values by $\delta Q_1$ and $\delta Q_2$. The resulting energy change $E_1^+(n) = F(n+1, \xi+1) - F(n, \xi)$ is given by

$$E_1^+(n) = Q_1 \frac{\delta Q_1}{C_1} + Q_2 \frac{\delta Q_2}{C_2} + \frac{(\delta Q_1)^2}{2C_1} + \frac{(\delta Q_2)^2}{2C_2} - (\delta Q_1 - e)V. \qquad (4.70)$$

The charge deviation $\delta Q_i$ is determined from the charge conservation $\delta Q_2 - \delta Q_1 = e$ and Kirchhoff's law $(\delta Q_1/C_1) + (\delta Q_2/C_2) = 0$ as

$$\delta Q_1 = (C_1/C)e = (C_T/C_2)e, \qquad \delta Q_2 = -(C_2/C)e = -(C_T/C_1)e, \quad (4.71)$$

where $C = C_1 + C_2$ is the total capacitance of the island and $C_T = (1/C_1 + 1/C_2)^{-1}$ is the total capacitance of the series. Note that the charge deviations are independent of the number of excess electrons $n$. Substituting Eq. (4.71) into Eq. (4.70) and using Eq. (4.68), we have

$$E_1^+(n) = (1 + 2n)\frac{e^2}{2C} + \frac{C_2}{C}eV. \qquad (4.72)$$

The energy changes for the other tunneling processes, $E_1^-(n) = F(n-1, \xi - 1) - F(n, \xi)$ and $E_2^\pm(n) = F(n \pm 1, \xi) - F(n, \xi)$, are calculated in the same way. The results are summarized as

$$E_1^\pm(n) = (1 \pm 2n)\frac{e^2}{2C} \pm \frac{C_2}{C}eV, \qquad E_2^\pm(n) = (1 \pm 2n)\frac{e^2}{2C} \mp \frac{C_1}{C}eV. \quad (4.73)$$

Using the golden rule formula for the tunnel Hamiltonian, the tunneling rate is calculated as [35]

$$\Gamma_i^\pm(n) = \frac{1}{e^2 R_i} \int_{-\infty}^{\infty} d\epsilon_\mathbf{k} \int_{-\infty}^{\infty} d\epsilon_\mathbf{p} f(\epsilon_\mathbf{k}) \left[1 - f(\epsilon_\mathbf{p})\right] \delta(\epsilon_\mathbf{k} - \epsilon_\mathbf{p} - E_k^\pm(n))$$

$$= \frac{1}{e^2 R_i} \frac{E_i^\pm(n)}{\left[\exp(E_i^\pm(n)/k_B T) - 1\right]}. \tag{4.74}$$

We assume the electrostatic equilibrium during the successive tunneling processes in the double junction and neglect correlations between different tunneling processes. The probability $p(n)$ for finding $n$ excess electrons in the island is determined by the tunneling rates for leaving this state to the states $n \pm 1$ and for coming into this state from the states $n \pm 1$, and satisfies the master equation,

$$\begin{aligned}
\frac{dp(n)}{dt} &= p(n-1)\Gamma(n-1 \to n) + p(n+1)\Gamma(n+1 \to n) \\
&\quad - p(n)\left[\Gamma(n \to n+1) + \Gamma(n \to n-1)\right],
\end{aligned} \tag{4.75}$$

where $\Gamma(n \to n \pm 1)$ is the rate for the transition $n \to n \pm 1$ of the charge state:

$$\Gamma(n \to n+1) = \Gamma_1^+(n) + \Gamma_2^+(n), \qquad \Gamma(n \to n-1) = \Gamma_1^-(n) + \Gamma_2^-(n).$$

We calculate the stationary probability $p(n)$ by requiring $dp(n)/dt = 0$, for which $p(n)$ satisfies the condition of detailed balancing

$$p(n)\left[\Gamma_1^+(n) + \Gamma_2^+(n)\right] = p(n+1)\left[\Gamma_1^-(n+1) + \Gamma_2^-(n+1)\right], \tag{4.76}$$

with the normalization condition $\sum_{n=-\infty}^{\infty} p(n) = 1$. Knowing the stationary probability $p(n)$ from the master equation, we can calculate the tunnel current through junctions 1 and 2

$$I = -e \sum_{n=-\infty}^{\infty} p(n)\left[\Gamma_1^+(n) - \Gamma_1^-(n)\right] = -e \sum_{n=-\infty}^{\infty} p(n)\left[\Gamma_2^+(n) - \Gamma_2^-(n)\right]. \tag{4.77}$$

The equality of the second and third terms is ensured by the detailed balance condition (4.76). In the next subsection, we examine the *I-V* characteristic for a metal-droplet system in detail by using the Eq. (4.77).

### 4.3.1.1 Coulomb blockade and Coulomb staircase

Figures 4.12(a) and (b) show the tunnel current $I$ and the probability $p(n)$ at $T = 0$ as functions of bias voltage $V$ for double junction systems with equal resistance $R_1 = R_2$ and with very different resistances $R_1 = 100R_2$, respectively. In both systems, the capacitance is taken to be $C_1 = \sqrt{2}C_2$. As seen in Fig. 4.12, the electron tunneling is blocked at small bias-voltages below the threshold voltage $V_{th}$ ($V < V_{th}$), where the tunnel current vanishes and the island charge remains to be zero. In this CB region, the neutral charge state ($n = 0$) is stable because all transitions to the states of $n \neq 0$ increase the electrostatic energy by $E_i^\pm(0)$. This condition is expressed by $E_i^\pm(0) > 0$, for

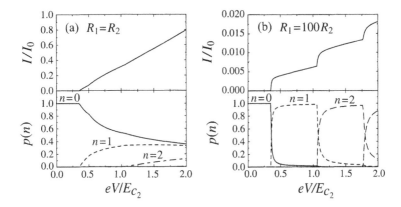

Figure 4.12: (a) Tunneling current $I$ and the probability $p(n)$ for $R_1 = R_2$ are plotted as functions of bias voltage at $T = 0$ in the upper and lower panels, respectively. (b) The same plot for $R_1 = 100R_2$ with a bottleneck of the tunnel current at the first junction. The bias voltage is normalized by $E_{C_2} = e^2/C_2$ and the tunnel current is normalized by $I_0 = e/(C_2 R_2)$.

which all tunneling rates vanish: $\Gamma_i^{\pm}(0) = 0$. Therefore, the threshold voltage for the CB is obtained from Eq. (4.73) as

$$V_{th} = \min\left\{ \frac{e}{2C_1}, \frac{e}{2C_2} \right\}, \qquad (4.78)$$

indicating that $V_{th}$ is determined by the larger one of the junction capacitances. In the case of $C_1 > C_2$ in Fig. 4.12, the threshold voltage is given by $V_{th} = e/2C_1$ or in the normalized form $eV_{th}/E_{C_2} = (C_2/C_1)/2 = \sqrt{2}/4$ with $E_{C_2} = e^2/C_2$.

Above $V_{th}$, one finds a clear difference in the $I$-$V$ curves of the symmetric and highly asymmetric junctions. The step-like structure called "Coulomb staircase" (CS) [35-39] appears in highly asymmetric junctions. This is because the 1st junction behaves as a bottleneck of the tunnel current for $R_1 \gg R_2$ and the bottleneck brings about the CS. Once the bias voltage exceeds the threshold voltage, an electron is allowed to tunnel onto the island through the 2nd junction because of $E_2^+(0) < 0$, and a new charge state of $n = 1$ becomes stable in the voltage region of the 1st step-like structure ($e/2C_1 < V < 3e/2C_1$). The island is charged through the 2nd junction until the number of excess electrons reaches its maximum value $n_{\max} = 1$. Occasionally, an electron tunnels through the 1st junction resulting in a tunnel current through the double junction. In that case, the probability for the charge state $n = 1$ is very close to unity in contrast to that for the symmetric junction as shown in the lower panels of Figs. 4.12(a) and (b). As the bias voltage increases, the maximum number of excess electrons $n_{\max}$ changes its value and the tunnel current shows jumps at $V = (n_{\max} + \frac{1}{2})e/C_1$. The tunnel current in Eq. (4.77) is approximated as $I = -E_1^-(n)/eR_1$ by using the $T = 0$ limit of Eq. (4.74) for negative $E_1^-(n)$, and $p(n) \sim 1$, so that the $I$-$V$ curve and its slope are given by

$$I = \frac{1}{R_1}\left[\frac{e}{C}n_{\max} - \frac{e}{2C} + \frac{C_2}{C}V\right], \qquad \frac{dI}{dV} \approx \frac{1}{R_1}\frac{C_2}{C}, \qquad (4.79)$$

reflecting the fact that the current is limited by the larger resistance $R_1$. Moreover, the $I$-$V$ curve has an abrupt increase in current by $\Delta I = e/(R_1C)$ at $V = (n_{\max} + \frac{1}{2})e/C_1$. The maximum number $n_{\max}$ in Eq. (4.79) is determined by the tunneling process through the 2nd junction as, $E_2^+(n_{\max}) > 0$ and $E_2^+(n_{\max} - 1) \leq 0$, which is written as

$$\frac{C_1}{e}V - \frac{1}{2} < n_{\max} \leq \frac{C_1}{e}V + \frac{1}{2}. \qquad (4.80)$$

Therefore, the period of the CS is determined by the capacitance at the 1st junction though it is determined by the tunneling process through the 2nd junction.

### 4.3.1.2   Coulomb staircase and Tunnel magnetoresistance

In the above subsections, we have neglected the spin-degree of freedom. However, electrons have spin as well as charge. In the magnetic tunnel junctions, the tunnel current depends on the relative orientation of magnetizations in the ferromagnetic electrodes. In magnetic double tunnel junctions with different tunnel resistances, it has been predicted that the TMR oscillates with respect to the bias voltage [40, 41]. Experimentally, the TMR effect has been observed in the vacuum tunneling of spin-polarized electrons between a Fe-coated W tip and a Gd island [42].

We consider a double tunnel junction consisting of a non-magnetic STM-tip, a magnetic metal-droplet and a magnetic base electrode shown in Figs. 4.13(a) and (b). The tunnel resistance $R_i$ at $i$-th junction is given in terms of the tunnel resistance $R_{i\sigma}$ of spin-$\sigma$ channel by $1/R_i = 1/R_{i\uparrow} + 1/R_{i\downarrow}$. The tunnel resistance in the first junction is spin-independent because the tip is non-magnetic, while the tunnel resistance in the second junction depends on the relative orientation of the magnetization vectors between the droplet and the base electrode. For simplicity, we neglect the spin accumulation in the droplet and assume that the energy changes due to the electron tunneling are the same for spin-up and spin-down electrons. Therefore the period of the CS is not affected by the spin. However, the tunneling rates depend on the spin through the junction resistance as

$$\Gamma_{i\sigma}^{\pm}(n) = \left(\frac{1}{e^2 R_{i\sigma}}\right)\frac{E_i^{\pm}(n)}{\exp[E_i^{\pm}(n)/k_BT] - 1}. \qquad (\sigma = \uparrow, \downarrow) \qquad (4.81)$$

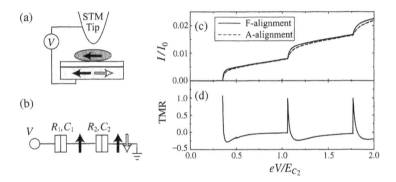

Figure 4.13: (a) Schematic diagram of the tunnel junction consisting of an non-magnetic STM-tip, a magnetic-metal droplet and a magnetic base electrode. Arrows indicate the magnetizations of magnetic-metals. In the base electrode, Solid(open) arrow represents the Ferromagnetic (Antiferromagnetic) alignment. (b) The corresponding theoretical model. (c) Tunneling currents for F and A alignments are plotted in the upper panel by the solid and dotted lines, respectively. The bias voltage is normalized by $E_{C_2} = e^2/C_2$ and the tunnel current is normalized by $I_0 = E_{C_2}/eR_{2\downarrow}$. (d) The corresponding TMR.

The tunnel current is calculated from Eqs. (4.74)-(4.77).

What is the difference between non-magnetic and ferromagnetic tunnel junctions? We consider ferromagnetic (F) and antiferromagnetic (A) alignments of the magnetizations. For the F (A) alignment, the magnetization vectors are set to be parallel (antiparallel) as shown in Figs. 4.13(a) and (b). In Fig. 4.13(c), we plot the tunnel current in the F and A alignments by the solid and dotted lines, respectively. The tunnel resistances are taken to be $R_{1\uparrow}:R_{1\downarrow}:R_{2\uparrow}^F:R_{2\downarrow}^F = 400:100:16:1$, and $R_{2\uparrow}^A = R_{2\downarrow}^A = (R_{2\uparrow}^F R_{2\downarrow}^F)^{1/2}$, which corresponds to the resistance symmetry $R_1 = 40R_2^A = (15/8)R_2^F$ and the spin polarization $P_1 = P_2 = 60\%$ of the ferromagnets in the droplet and base electrode. The junction capacitances are taken to be $C_1 = \sqrt{2}C_2$ and the temperature is zero. We see that the period of the CS is the same for both alignment since the energy changes are independent of spin. However, the magnitude of the currents between the F and A alignments are different, especially near the thresholds $V_{th}(n)$; the $I$-$V$ curve of the F alignment increases more sharply than that of the A alignment as seen in Fig. 4.13(c). This is because the magnitude of the tunnel current across the first junction is independent of the direction of the magnetization in the base electrode, while that in the second junction depends on the direction, making the tunnel resistances at junctions 1 and 2 more asymmetric in the F alignment than in the A alignment: $R_1/R_2^F > R_1/R_2^A$. The difference between the $I$-$V$ curves of the F and A alignments brings about the TMR oscillation with respect to the bias voltage as shown in Fig. 4.13(d). The TMR has sharp peaks at the bias voltages where the tunnel current jumps. The period of the oscillation is the same as that of the CS.

## 4.3.2  Coulomb staircase and TMR in STM current through magnetic granular films

Scanning tunneling microscopy (STM) and spectroscopy (STS) are useful for investigating the single electron tunneling in granular films. Recently, the CS has been observed in the $I$-$V$ curve in highly resistive granular films using STM [43-45]. In this section, we consider a system consisting of a non-magnetic STM-tip and a magnetic granular film, and discuss the CS in thin and thick granular films. In contrast with the droplet system in Sec. 4.3.1, the physics behind the CS in a granular film is not clear because the STM current passes through many granules of different size and may spread out to form a 3D network in a thick granular film. In this system, the tunnel current between the tip and a granule on the surface can be varied by changing the distance between them. When the tunnel resistance between the tip and the granule is much larger than those between granules in the interior of the film, a bottleneck of the tunnel current is created between the tip and the granule. As a consequence, the CS appears in the $I$-$V$ curve and has a single period of step structure. The period of CS does not depend on the thickness of granular films, but depends solely on the capacitance between the STM tip and the granule probed by the tip. It is shown that the experimental results on 10 nm and 1 $\mu$m thick Co-Al-O granular films at room temperature are well explained by a model of the tunnel junction array with a bottleneck. The oscillation of the MR with the same period of CS is predicted.

### 4.3.2.1  Thin granular films

We first consider a thin granular film which contains a few granules between the tip and the base electrode. In Fig. 4.14, the experimental setup and the corresponding theoretical model, i.e., 1D array of $N$ junctions, are shown schematically. Each junction is characterized by the tunnel resistance $R_j$ and the capacitance $C_j$, and carries the charge $Q_j$. The charges on the granules is represented by the set of numbers $\{n_i\} = (n_1, n_2, \cdots n_{N-1})$ with $n_j = (Q_j - Q_{j+1})/e$. We neglect the spin accumulation and define the tunnel resistances as follows: $1/R_j \equiv 1/R_{j\uparrow} + 1/R_{j\downarrow}$.

A basic formalism is constructed in the same procedure as in Sec. 4.3.1. For the 1D array of $N$ junctions, it is convenient to consider the energy change for leftward and rightward tunneling of an electron through the $l$-th junction.

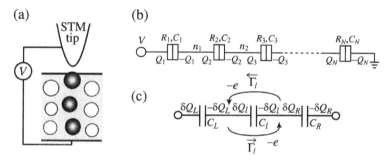

Figure 4.14: (a) A system with an STM tip and a thin granular film is schematically shown. The solid (open) circles represent granules in (out of) the conducting path. The insulating matrix is indicated by the shading. (b) Equivalent circuit of the system. Each junction is characterized by the tunnel resistance $R_j$ and the capacitance $C_j$, and carries the charge $Q_j$. (c) Reduced equivalent circuit (see text).

The energy change is expressed in terms of the charge $Q_j$ and its deviation $\delta Q_j$ as

$$\overset{\leftrightarrows}{E}_l(\{n_i\}) = \sum_{j=1}^{N} Q_j \frac{\delta Q_j}{C_j} + \sum_{j=1}^{N} \frac{(\delta Q_j)^2}{2C_j} - (\delta Q_1 \pm e\delta_{1,l})V, \qquad (4.82)$$

where $\{n_i\} = (n_1, n_2, \ldots, n_{N-1})$ represents the initial charge state on the granules, the superscript $\leftarrow$ ($\rightarrow$) denotes the leftward (rightward) tunneling, $N$ is the number of tunnel junctions, and $\delta_{1,l}$ is Kronecker's delta. Making use of the relation

$$\sum_j Q_j \frac{\delta Q_j}{C_j} = \sum_i (Q_i - Q_{i+1}) \sum_j \frac{\delta Q_j}{C_j} = \sum_{i=1}^{N-1} n_i e \left( \sum_{j=1}^{i} \frac{\delta Q_j}{C_j} \right),$$

the energy change $\overset{\leftrightarrows}{E}_l(\{n_i\})$ is expressed in terms of $\delta Q_i$ and $n_i$:

$$\overset{\leftrightarrows}{E}_l(\{n_i\}) = \sum_i \left( \sum_{j<i} \frac{\delta Q_j}{C_j} \right) n_i + \frac{1}{2} \sum_i \frac{(\delta Q_i)^2}{C_i} - (\delta Q_1 \pm e\delta_{1,l})V. \qquad (4.83)$$

The deviation $\delta Q_i$ is determined by Kirchhoff's law by considering the equivalent network shown in the inset of Fig. 4.14(b), where $1/C_L = \sum_{i<l}(1/C_i)$ and $1/C_R = \sum_{i>l}(1/C_i)$. When an electron tunnels from right (left) to left (right) through the $l$-th junction, $\delta Q_i$ satisfy the following equations: $\delta Q_i \equiv \delta Q_L$ for $i < l$, $\delta Q_i \equiv \delta Q_R$ for $i > l$, $-\delta Q_L + \delta Q_l = \mp e$ and $-\delta Q_l + \delta Q_R = \pm e$, and $(\delta Q_L/C_L) + (\delta Q_l/C_l) + (\delta Q_R/C_R) = 0$. After some algebras, we have

$$\delta Q_l = \mp e\left[1 - (C_T/C_l)\right], \qquad \delta Q_L = \delta Q_R = \pm e(C_T/C_l), \qquad (4.84)$$

where $C_T$ is the total capacitance of the array: $1/C_T = \sum_{i=1}^{N}(1/C_i)$. Substituting Eq. (4.84) into Eq. (4.83), we have the energy change,

$$
\begin{aligned}
\overleftrightarrow{E}_l(n_1,\ldots,n_{N-1}) \\
= \frac{C_T}{C_l}\Bigg[ \pm e^2 \Bigg\{ \sum_{i=1}^{l-1}\left(\sum_{j=1}^{i}\frac{1}{C_j}\right)n_i - \sum_{i=l}^{N-1}\left(\frac{1}{C_T}-\sum_{j=1}^{i}\frac{1}{C_j}\right)n_i \Bigg\} \\
+ \frac{e^2}{2}\left(\frac{1}{C_T}-\frac{1}{C_l}\right)\mp eV \Bigg].
\end{aligned}
\tag{4.85}
$$

It is easily checked that $\overleftrightarrow{E}_l(n_i)$ reduces to Eq. (4.73) for $N = 2$ and satisfies the relation $\overleftarrow{E}_l(\ldots,n_{l-1},n_l,\ldots) + \overrightarrow{E}_l(\ldots,n_{l-1}+1,n_l-1,\ldots) = 0$. The tunneling rate from the initial state with $\{n_i\}$ is given by

$$
\overleftrightarrow{\Gamma}_l(\{n_i\}) = \left(\frac{1}{e^2 R_l}\right)\frac{\overleftrightarrow{E}_l(\{n_i\})}{\exp\left[\overleftrightarrow{E}_l(\{n_i\})/T\right]-1}.
\tag{4.86}
$$

To obtain the tunneling current, we construct the master equation for the probability of states $p(\{n_i\}_\alpha)$ [35, 46, 47], which is given in the matrix form by

$$
\dot{\boldsymbol{p}} = \mathsf{M}\boldsymbol{p}(t),
\tag{4.87}
$$

where $\boldsymbol{p} = (\ldots,p(\{n_i\}_\alpha),\ldots)^T$ and $\mathsf{M}$ is the transition matrix in the configuration space constructed by $\{n_i\}_\alpha \equiv (n_1,\ldots,n_{N-1})$ with index $\alpha$ labeling the different charge states. The matrix $\mathsf{M}$ is expressed as,

$$
\mathsf{M}_{\{n_i\}_\alpha,\{n_i\}_\beta} = \begin{cases} \Gamma(\{n_i\}_\alpha \to \{n_i\}_\beta) & \text{for } \alpha \neq \beta \\ -\sum_\gamma \Gamma(\{n_i\}_\alpha \to \{n_i\}_\gamma) & \text{for } \alpha = \beta \end{cases},
\tag{4.88}
$$

where $\Gamma(\{n_i\}_\alpha \to \{n_i\}_\beta)$ represents the tunneling rate from the initial charge state $\{n_i\}_\alpha$ to the final one $\{n_i\}_\beta$. For double tunnel junctions with $N = 2$, Eq. (4.87) is reduced to Eq. (4.75). The transition matrix $\mathsf{M}$ has the eigenvalue of zero whose eigenvector represents the stationary state. By using the solution $p(\{n_i\}_\alpha)$, the tunneling current through the $l$-th junction is given by

$$
I_l = e\sum_\alpha p(\{n_i\}_\alpha)\left[\overleftarrow{\Gamma}_l(\{n_i\}_\alpha) - \overrightarrow{\Gamma}_l(\{n_i\}_\alpha)\right],
\tag{4.89}
$$

where $\overleftarrow{\Gamma}_l(\{n_i\}_\alpha)$ $(\overrightarrow{\Gamma}_l(\{n_i\}_\alpha))$ is the tunneling rate through the $l$-th junction from right (left) to left (right) with the initial charge state $\{n_i\}_\alpha$. The tunneling current $I_l$ in Eq. (4.89) at the $l$-th junction is the same for all $l = 1,\ldots,N$.

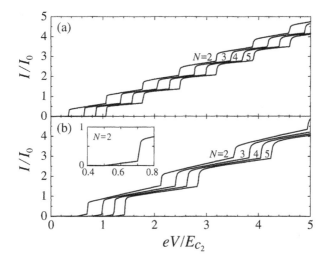

Figure 4.15: Tunneling current normalized with $I_0 = E_{C_2}/eR_1$ is plotted as a function of the bias voltage at $T = 0$, where $E_{C_2} = e^2/C_2$. The number of junctions are $N = 2 \sim 5$ from left to right. The tunnel resistance ratio is taken to be $R_1:R_2:R_3:R_4:R_5 = 1000:1:1:1:1$. The capacitance ratio is (a) $C_1:C_2:C_3:C_4:C_5 = \sqrt{2}:1:\sqrt{3}:\sqrt{5}:\sqrt{6}$, and (b) $C_1:C_2:C_3:C_4:C_5 = 1/\sqrt{2}:1:\sqrt{3}:\sqrt{5}:\sqrt{6}$. The tunneling current with $n_1 = 0$ as a dominant state for $N = 2$ is shown in the inset of panel (b).

We first discuss the *I-V* characteristics for a 1D array of tunnel junctions with a bottleneck of the tunneling current between the tip and a granule on the surface. In Fig. 4.15, the numerical results are shown for the tunnel junctions with $N = 2 \sim 5$ at $T = 0$. It is seen that single electron tunneling is blocked as long as the bias voltage $V$ is lower than the threshold value $V_{th}$ [48], and that $V_{th}$ increases with increasing $N$. Let us suppose that an electron tunnels through the $l$-th junction. Then, from Eq. (4.85), the energy change due to the tunneling is given by $\overset{\leftrightarrows}{E_l}(0,\dots,0) = (e^2/2)(C_T^{-1} - C_l^{-1}) - eV$. This leads to a threshold voltage for the transition, $V_{th}^l = (e/2)(C_T^{-1} - C_l^{-1})$, above which the initial state $(0,\dots,0)$ is unstable, i.e., $\overset{\leftarrow}{E_l}(0,\dots,0) < 0$. The minimum of $V_{th}^1,\dots,V_{th}^N$, i.e., $V_{th} = \min_l\{V_{th}^l\}$, yields the threshold voltage

$$V_{th} = \frac{e}{2}\left(\frac{1}{C_T} - \frac{1}{\min_l\{C_l\}}\right). \tag{4.90}$$

When $N$ is large and the capacitances are nearly the same, the total capacitance $C_T$ of the 1D array is inversely proportional to the number of granules in the array, so that the threshold voltage becomes $V_{th} \sim (e/2\langle C\rangle)(N-1)$, where $\langle C\rangle$ is the average capacitance. This indicates that $V_{th}$ increases in proportion to the number of granules [48, 49] and thus film thickness as shown in Fig. 4.15.

The CS is classified into two types as shown in Figs. 4.15(a) and (b), depending on whether the capacitance $C_1$ of the bottleneck is the smallest or not. In the case that $C_1$ is not the smallest and $V_{th} = \min_{l \neq 1}\{V_{th}^l\}$, a typical example of CS is shown in Fig. 4.15(a), where electrons start to accumulate at the bottleneck once the bias voltage exceeds $V_{th}$. Until the accumulated electrons tunnel out through the bottleneck, the voltage drop caused by them forbids electrons to tunnel through the other junctions. Therefore, the stable state is given by $(n_1, 0, \ldots, 0)$ and the tunneling current jumps at the bias voltage where the number of accumulated electrons $n_1$ changes. This number $n_1$ is the minimum value satisfying the conditions

$$\overleftarrow{E}_1(n_1, 0, \ldots, 0) < 0, \qquad \overleftarrow{E}_{l \neq 1}(n_1, 0, \ldots, 0) \geq 0. \qquad (4.91)$$

The first condition represents that accumulated electrons tunnel out through the first junction. The second indicates that electrons cannot tunnel through the other junctions, and is rewritten as $(e^2/C_1)n_1 \geq eV - V_{th}^{l \neq 1}$. Therefore, the tunneling current jumps at the bias voltage $V$ given by

$$V = \min_{l \neq 1} V_{th}^l + (e^2/C_1)(n_1 - 1), \qquad (4.92)$$

indicating that the CS has a single period of $e/C_1$. The tunneling current for each plateau of the CS is approximately given by $I \simeq e\overleftarrow{\Gamma}_1(n_1, 0, \ldots, 0)$.

On the other hand, when $C_1$ is the smallest, the tunneling current does not jump at $V_{th}$ as shown in Fig. 4.15(b). For $V_{th} < V < \min_{l \neq 1} V_{th}^l$, the tunneling current is approximately given by $I \simeq e\overleftarrow{\Gamma}_1(0, \ldots, 0)$. Once $V$ exceeds $\min_{l \neq 1} V_{th}^l$, electrons start to accumulate at the bottleneck and the $I$-$V$ curve shows a CS. The bias voltage $V$ at which the tunneling current jumps is given by Eq. (4.92).

When the bottleneck is placed at another junction $j(1 < j < N)$, the stable state is $(0, \ldots, 0, n_{j-1} = -n_j, n_j, 0, \ldots, 0)$ and the period of the CS is determined by the capacitance $C_j$ at the bottleneck. The CS is classified into two types in the same way as those shown in Figs. 4.15(a) and (b). However, the criterion is whether the capacitance $C_j$ is the smallest or not.

The experimental $I$-$V$ curves for a 10-nm thick $Co_{36}Al_{22}O_{42}$ film are plotted in Fig. 4.16(a). Samples were prepared by an oxygen-reactive sputtering with a Co-Al alloy target and there exist two or three granules between the tip and base electrode. A conventional STM system with a Pt tip was operated at room temperature under high vacuum ($\sim 10^{-8}$ Torr). The $I$-$V$ curves were obtained by placing the tip on a Co granule. Even at room temperature, the tunneling current shows a clear CS with a single period. The tunnel resistance between granules for Co-Al-O granular films is estimated to be $10^5 \sim 10^6$ $\Omega$ from the average diameter of granules ($\sim 3$ nm), the average intergranular distance ($\sim 1$ nm), and electrical resistivity [45, 50]. On the other hand, the tunnel

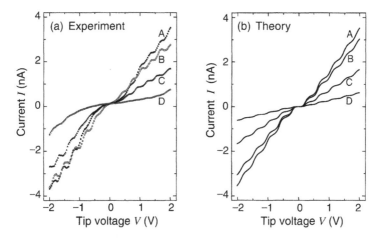

Figure 4.16: (a) Experimental $I$-$V$ curves for a 10-nm thick $Co_{36}Al_{22}O_{42}$ measured at room temperature. A, B, C and D refer to different distance between the STM tip and the surface of the sample. The lateral position for A and B is different from that for C and D. (b) Corresponding theoretical curves in a triple tunnel junction system at $T = 300$ K. The tunnel resistance at the bottleneck is taken to be $R_1 = 600, 700, 1300$, and $3500$ MΩ for the line A, B, C and D, respectively. The other tunnel resistances are $R_2 = R_3 = 1$ MΩ and the capacitances are $C_1 = 4.48 \times 10^{-19}$ F, $C_2 = 2.13 \times 10^{-19}$ F and $C_3 = 3.62 \times 10^{-19}$ F for all curves.

resistance between the tip and a granule on the surface, $R_1$, is $10^8 \sim 10^9$ Ω. Therefore, $R_1$ is $10^2 \sim 10^4$ times larger than the other tunnel resistances.

The calculated $I$-$V$ curves are shown in Fig. 4.16(b). A triple tunnel junction model with a bottleneck between the tip and a granule on the surface is used for the calculation. Parameter values were chosen in the range estimated from the experiments. We find that the theoretical curves explain the experimental ones very well.

We also study the TMR in the system with a non-magnetic STM-tip and a magnetic granular film containing ferromagnetic metal granules. Since the master equation depends sensitively on the relative orientation of magnetizations of the granules, especially around the step in the CS, a strong modulation of the TMR is expected. The magnetic property of granules is characterized by their spin polarization $P$. We use the value $P = 0.35$ of Co. In the triple junction system shown in Figs. 4.17(a) and (b), the tunnel resistance $R_2$ depends on the relative angle $\theta$ between magnetizations of two granules as $R_2 \sim 1/(1 + P^2 \langle \cos \theta \rangle)$ [51]. The average $\langle \cos \theta \rangle$ can be expressed by using the relative magnetization $m$ of the system as $\langle \cos \theta \rangle = m^2$. Since the granules of nanometer-size are in superparamagnetic state at room temperature, the

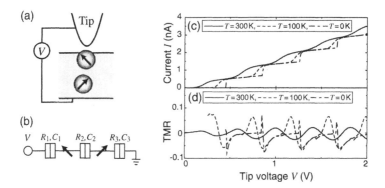

Figure 4.17: (a) Schematic of the tunnel junction consisting of the non-magnetic STM-tip, magnetic granular film and non-magnetic base electrode. The arrows indicate the magnetizations of magnetic granules. (b) Equivalent circuit of corresponding theoretical model. (c) Current-voltage curves at zero magnetic field ($m = 0$) at $T = 0\,\mathrm{K}$ (solid), $100\,\mathrm{K}$ (dotted), and $300\,\mathrm{K}$ (dot-dashed). The junction resistances and capacitances are the same as those used for the curve A in Fig. 4.16. (d) The corresponding TMR.

relative magnetization is $m = 0$ in the absence of the magnetic field. When an applied magnetic field is strong enough to align all the magnetization vectors parallel, the relative magnetization is $m = 1$. The TMR is defined as $[G(1) - G(0)]/G(0)$, where $G(m)$ is the differential conductance of the STM current in the system of relative magnetization $m$. Figures 4.17(c) and (d) show the tunneling current and corresponding TMR, respectively. The solid, dotted and dot-dashed lines denote those at $T = 0$, 100 and $300\,\mathrm{K}$, respectively. The junction resistances and capacitances are the same as those used for the curve A in Fig. 4.16. The TMR oscillates with the same period as the CS, which is determined by the capacitance between the tip and a granule on the surface. Sharp peaks appear at the step edge of the CS at $T = 0$, and becomes broad and shifts to the lower bias voltage as temperature increases. Experimentally, the TMR has not yet been reported since it is difficult to fix the position of the tip and measure the $I$-$V$ characteristics when the magnetic field is applied.

### 4.3.2.2   Thick granular films

The $I$-$V$ curves for a 1-$\mu$m thick $\mathrm{Co_{36}Al_{22}O_{42}}$ film have been measured experimentally. The results are plotted in Fig. 4.18(a). We see that the tunnel current varies stepwise with a constant interval with respect to $V$, showing the CS with a single period. This feature is essentially the same as in the thin film Fig. 4.16, despite the fact that a few granules are involved in tunneling in

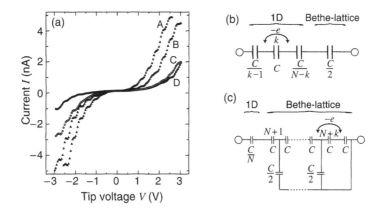

Figure 4.18: (a) Experimental *I-V* curves for a $1$-$\mu$m thick $Co_{36}Al_{22}O_{42}$ film measured at room temperature. A, B, C and D refer to different distance of the STM tip from the surface of the sample. The lateral tip position for A and B is different from that for C and D. (b) The equivalent circuit for the electron tunneling in the 1D array, where the Bethe-lattice network is replaced by its total capacitance $C/2$. (c) The equivalent circuit for the electron tunneling in the Bethe-lattice network.

the thin film, while 200 to 300 Co granules in the thick film. The threshold voltage of the thick film is $V_{th} \simeq 0.5\,\mathrm{V}$, which is much smaller than the value expected in the 1D array model, which predicts a much large value because $V_{th}$ is proportional to the number of granules along the conduction path; $V_{th}$ of $1$-$\mu$m thick film should be about 100 times larger than that of the 10-nm thick film.

The discrepancy can be resolved by considering a network of the conducting paths in a thick granular film between an STM-tip and the base electrode shown in Figs. 4.19(a) and (b). In the thick granular film, the conducting path branches out to form a 3D network inside the film as shown in Figs. 4.19(a) and (b). In order to describe the network of the conducting paths in a thick film containing many granules in the direction perpendicular to the film plane between the tip and the substrate, we introduce a theoretical model of 1D array connected to a Bethe-lattice network shown in Fig. 4.19(b). In this junction array, the energy change due to the single electron tunneling through the *l*-th junction is given by

$$\overline{\overline{E}}_l(\{n_i\}) = \sum_i \left( \sum_{j<i} \frac{\widetilde{\delta Q_j^l}}{C_j} \right) n_i + \frac{1}{2} \sum_i \frac{\left( \delta Q_i^l \right)^2}{C_i} - (\delta Q_1^l \pm e\delta_{1,l})V, \qquad (4.93)$$

where $\widetilde{\sum}_i$ represents the summation along the conducting path. The first term of Eq. (4.83) is different from that of (4.93) because the relation between the

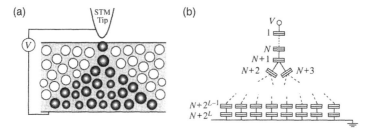

Figure 4.19: (a) Schematic drawing of the system with an STM tip and a thick granular film. The filled circles represent granules in the conducting paths. The insulating matrix is indicated by the shading. (b) The corresponding theoretical model. Each junction is characterized by the tunnel resistance $R_j$ and the capacitance $C_j$, and carries the charge $Q_j$.

number of excess electrons is given by $n_{N+j} = Q_{N+j} - Q_{N+2j} - Q_{N+2J+1}$ in the Bethe-lattice network.

Because the bottleneck is placed between the tip and a granule on the surface, the stable states are given by $(n_1, 0, \ldots, 0)$ and the tunneling current jumps when the number of accumulated electrons $n_1$ changes. The period of CS in the system with a thick film shown in Fig. 4.18(a) is also determined from Eqs. (4.83) and (4.91), as long as the tip is coupled to a single granule on the surface and the bottleneck is created between them. Therefore, the voltage at which the tunneling current jumps is given by Eq. (4.92). The CS has a single period $e/C_1$ determined by the capacitance at the bottleneck, even for a thick film. The equivalent network used to obtain the threshold voltage for the $l$-th junction $V_{th}^l$ is shown in Figs. 4.18(b) and (c), where we assume, for simplicity, the same capacitance $C$ for all junctions. The key point is that for electron tunneling in the 1D array, the Bethe-lattice network is replaced by its total capacitance as shown in Fig. 4.18(b). The total capacitance for the Bethe-lattice is $C/2$ and the bias voltage $V$ at which the tunneling current jumps (see Eq. (4.92)) is given by $V = (e/2C)(N + 1) + (e/C)(n_1 - 1)$. The experimental $I$-$V$ curves shown in Fig. 4.18(a) are consistent with our model with $N \simeq 2$. The TMR oscillation with the same period as CS is expected for a system with a thick granular film. The period is determined by the capacitance of the bottleneck though a thick granular film contains many granules in the direction perpendicular to the film plane.

### 4.3.3   Higher-order tunneling and Magnetoresistance in Double tunnel junctions

A single electron transistor (SET) is the three-terminal device of a double tunnel junction consisting of a small island between bulk external electrodes and a gate capacitively coupled to the island. The circuit diagram is depicted

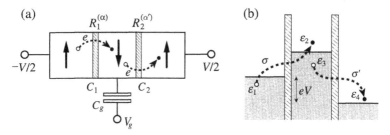

Figure 4.20: (a) Schematic diagram of a ferromagnetic single-electron transistor (SET) circuit with two tunnel junctions and a capacitive gate. The arrows indicate the magnetizations of the electrodes (left and right) and the island (middle). (b) Energy diagram illustrating tunneling processes in two junctions involved in inelastic cotunneling.

in Fig. 4.20. Electric charges flow through the device by tunneling driven by the transport voltage $V$. The value of the current is controlled by the gate voltage $V_g$.

Effects of CB on charge transport in metallic SETs have been extensively studied [52-58] It is well-known that there are two distinct tunneling processes by which electrons are transferred between the electrodes via a small island: sequential tunneling and cotunneling. In the sequential tunneling, there is no correlation between tunneling events onto and out of the island (see Fig. 4.20). In this process, electron tunneling in either of the two junctions causes an increase of charging energy, and is suppressed by CB. However, in the cotunneling, two electrons tunnel in a correlated fashion, i.e., an electron tunnels onto the island while a second electron simultaneously leaves the island through the other junction; the island is only virtually charged in this cotunneling process, and therefore there is no increase in charging energy in the overall tunneling process. While the cotunneling is a higher-order tunneling process, it contributes to a finite current in the CB regime [55, 59].

Single electron transistors consisting of ferromagnetic metals such as Ni/ NiO/Co/NiO/Ni [60, 61] and hybrid tunnel junctions containing ferromagnetic granules of nanometer size in the barrier [62-64] have recently been fabricated. In both systems, the enhanced TMR has been observed in the CB regime. A similar effect is also observed in highly resistive magnetic granular systems [50] as shown in the next section.

In this section, we discuss the effect of higher-order tunneling (cotunneling) in a small ferromagnetic double tunnel junction (FM/I/FM/I/FM). Since cotunneling is a fourth order process in the tunneling matrix elements, the tunnel current is sensitive to the relative orientation of magnetization between electrodes and island, and therefore the TMR is strongly enhanced by the CB [65].

#### 4.3.3.1   Charging energy for magnetic SET

As discussed in Sec. 4.3.1, when the island of the SET is small enough and the temperature and bias voltage are low enough, the electrostatic energy of excess electrons on the island has a significant effect on charge transport through the SET. By incorporating the effect of gate voltage into the formula derived in Sec. 4.3.1, one obtains the energy changes associated with forward tunneling of an electron through the 1st junction $(n \to n + 1)$ and that through the 2nd junction $(n \to n - 1)$:

$$E_1^+(n) \;=\; (1 + 2n)E_c - \left(\frac{C_g}{C_\Sigma}\right)eV_g - \left(\frac{C_2 + \frac{1}{2}C_g}{C_\Sigma}\right)eV, \qquad (4.94)$$

$$E_2^-(n) \;=\; (1 - 2n)E_c + \left(\frac{C_g}{C_\Sigma}\right)eV_g - \left(\frac{C_1 + \frac{1}{2}C_g}{C_\Sigma}\right)eV, \qquad (4.95)$$

where $n$ is the excess electrons on the island, $E_c = e^2/2C_\Sigma$ is the charging energy, $C_\Sigma = C_1 + C_2 + C_g$ is the capacitance of the island, and $C_1$, $C_2$, and $C_g$ are those of junctions 1, 2, and the gate, respectively. The energy changes for the backward tunneling are, respectively, given by

$$E_1^-(n) = E_2^-(n) + eV, \qquad E_2^+(n) = E_1^+(n) + eV. \qquad (4.96)$$

In ordinary tunnel junctions with negligible charging effect, individual electrons are unstable for the forward tunneling across the junction, and lower their energy by the bias potential $(-eV/2 < 0)$. In the SET, however, there is a voltage region where the charge state of island with $n$ excess electrons is stable with respect to tunneling across the first and second junctions for voltages that satisfy

$$\left(n - \frac{1}{2}\right)e < C_g V_g + \left(C_2 + \frac{1}{2}C_g\right)V < \left(n + \frac{1}{2}\right)e, \qquad (4.97)$$

$$\left(n - \frac{1}{2}\right)e < C_g V_g - \left(C_2 + \frac{1}{2}C_g\right)V < \left(n + \frac{1}{2}\right)e. \qquad (4.98)$$

These inequalities give the rhombic-shaped regions in the $V - V_g$ plane along the $V_g$ axis as shown in Fig. 4.21, each of which indicates that the island of SET is charged with a fixed number $n$ $(n = 0, \pm 1, \dots)$ of electrons. Inside these regions, tunneling across the junctions causes an increase of electrostatic energy, so that all transitions are suppressed by CB and no current flows through the junctions at zero temperature.

#### 4.3.3.2   Cotunneling process

It has been pointed out [59] that, in the CB regime, there is a higher-order process of tunneling through both junctions via a virtual intermediate state

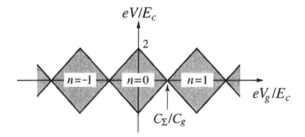

Figure 4.21: Stability diagram of an SET. The Coulomb blockade occurs inside the rhombic-shaped regions, each of which has a fixed number $n$ of electrons on the island. Outside these regions, the SET conducts.

with increased electrostatic energy. This *cotunneling* process is energetically favorable since the cotunneling lowers the electrostatic energy by $eV$, i.e., $E_1^+(n) + E_2^-(n+1) = E_2^+(n) + E_1^-(n-1) = -eV < 0$. Following Averin and Nazarov [35], we calculate the steady-state current through the double junctions

$$I(V) = \sum_{n=-\infty}^{\infty} p(n) \sum_{\sigma,\sigma'} \left[ \overrightarrow{\Gamma}_{1\sigma,2\sigma'}(n,V) - \overleftarrow{\Gamma}_{2\sigma',1\sigma}(n,V) \right], \qquad (4.99)$$

where $p(n)$ is the probability of charge state $ne$ on the island given in Eq. (4.73), $\overrightarrow{\Gamma}_{1\sigma,2\sigma'}(n,V)$ the forward tunneling rate at which an electron with spin $\sigma$ tunnels to the island through junction 1 and a different electron with spin $\sigma'$ tunnels out through junction 2, and $\overleftarrow{\Gamma}_{2\sigma',1\sigma}(n,V)$ the corresponding backward tunneling rate. The forward tunneling rate is

$$\begin{aligned}
\overrightarrow{\Gamma}_{1\sigma,2\sigma'}(n,V) = \ & \frac{\hbar}{2\pi e^4 R_{1\sigma}^{(\alpha)} R_{2\sigma'}^{(\alpha)}} \int_{-\infty}^{\infty} d\epsilon_1 d\epsilon_2 d\epsilon_3 d\epsilon_4 f\left(\epsilon_1\right) \left[1 - f(\epsilon_2)\right] \\
& \times f\left(\epsilon_3\right) \left[1 - f(\epsilon_4)\right] \delta(\epsilon_1 - \epsilon_2 + \epsilon_3 - \epsilon_4 + eV) \\
& \times \left| \frac{1}{\epsilon_2 - \epsilon_1 + E_1^+(n) + i\gamma_+} \right. \\
& \left. + \frac{1}{\epsilon_4 - \epsilon_3 + E_2^-(n) + i\gamma_-} \right|^2, \qquad (4.100)
\end{aligned}$$

where $R_{i\sigma}^{(\alpha)}$ is the tunnel resistance of junction $i$ for electrons with spin $\sigma$, and $\alpha$ represents the F (ferromagnetic) or A (antiferromagnetic) alignment of magnetization in the electrodes. The first term in $|\cdots|$ represents the process in which an electron tunnels to the island through the first junction and a different electron tunnels out through the second junction and the second term represents the reversed process, leaving an electron and a hole excitations on the island (see Fig. 4.20), and $\gamma_\pm$ is the decay rate of the charge state [65, 66]

$$\gamma_\pm = \frac{\hbar}{2} \sum_\sigma \left[\Gamma_{1\sigma}^\pm(n) + \Gamma_{2\sigma}^\pm(n)\right] = \frac{\hbar}{2e^2} \sum_{i=1}^2 \frac{E_i^\pm(n)}{R_i^{(\alpha)}} \coth \frac{E_i^\pm(n)}{2k_BT}, \qquad (4.101)$$

where $\Gamma_{i\sigma}^\pm(n)$ is given in Eq. (4.81) and $1/R_i^{(\alpha)} = 1/R_{i\uparrow}^{(\alpha)} + 1/R_{i\downarrow}^{(\alpha)}$. Note that the decay rate (4.101) behaves differently in the thermal regime ($k_BT \gg E_c$) and the quantum regime ($k_BT \ll E_c$):

$$\gamma_\pm = \frac{\hbar}{e^2}\left(\frac{1}{R_1^{(\alpha)}} + \frac{1}{R_2^{(\alpha)}}\right)k_BT, \qquad (k_BT \gg E_c),$$

$$\gamma_\pm = \frac{\hbar}{2e^2}\left(\frac{|E_1^\pm(n)|}{R_1^{(\alpha)}} + \frac{|E_2^\pm(n)|}{R_2^{(\alpha)}}\right), \qquad (k_BT \ll E_c).$$

By integrating Eq. (4.100) over spin and energy variables, we obtain

$$\sum_{\sigma,\sigma'}\overrightarrow{\Gamma}_{1\sigma,2\sigma'}(n,V) = \frac{(\hbar/2\pi e^4)}{R_1^{(\alpha)}R_2^{(\alpha)}} \int_{-\infty}^\infty d\epsilon \frac{\epsilon(\epsilon - eV)\exp\left[\frac{1}{2}\beta eV\right]}{\sinh\left[\frac{1}{2}\beta\epsilon\right]\sinh\left[\frac{1}{2}\beta(\epsilon - eV)\right]}|M|^2,$$

$$(4.102)$$

with $\beta = 1/(k_BT)$ and

$$M = \frac{1}{\epsilon + E_1^+(n) + i\gamma_+} + \frac{1}{-\epsilon + eV + E_2^-(n) + i\gamma_-}. \qquad (4.103)$$

Since the backward tunneling rate is related with the forward tunneling rate by the relation $\overleftarrow{\Gamma}_{2\sigma',1\sigma}(n,V) = \overrightarrow{\Gamma}_{1\sigma,2\sigma'}(n,-V) = \exp\left[-(eV/k_BT)\right]\overrightarrow{\Gamma}_{1\sigma,2\sigma'}(n,V)$, we have

$$I(V) = \left[1 - \exp\left(-\frac{eV}{k_BT}\right)\right]e \sum_{n=-\infty}^\infty p(n)\sum_{\sigma\sigma'}\overrightarrow{\Gamma}_{1\sigma,2\sigma'}(n,V). \qquad (4.104)$$

Equations (4.102)-(4.104) determine the current $I(V)$ through the junctions.

To extract an analytical behavior, we calculate the resistance $R_\alpha = dV/dI_\alpha$ at zero applied bias and gate voltages ($V = V_g = 0$). For simplicity, we consider the case where the initial charge of the island is zero, i.e., $n = 0$ ($p_0 = 1, p_{n\neq0} = 0$). The result is

$$R_\alpha^{-1} = \frac{2\hbar}{\pi e^2}\left[\frac{1}{R_1^{(\alpha)}R_2^{(\alpha)}}\right]\frac{1}{\beta}\int_{-\infty}^\infty d\epsilon \left[\frac{\beta\epsilon/2}{\sinh(\beta\epsilon/2)}\right]^2$$

$$\times \frac{E_c^2 + \gamma^2}{\left[(\epsilon - E_c)^2 + \gamma^2\right]\left[(\epsilon + E_c)^2 + \gamma^2\right]}, \qquad (4.105)$$

with $\gamma = (\hbar/2e^2)\left[(1/R_1^{(\alpha)}) + (1/R_2^{(\alpha)})\right] E_c \coth \frac{1}{2}\beta E_c$. At high temperatures ($k_B T \gg E_c$) where the charging effect plays no role, we may put $E_c = 0$ and calculate the integral in Eq. (4.105), yielding a series resistance

$$R_\alpha = R_1^{(\alpha)} + R_2^{(\alpha)}. \tag{4.106}$$

At low temperatures in the CB regime ($k_B T \ll E_c$), Eq. (4.105) reduces to

$$R_\alpha \approx \frac{3e^2}{4\pi\hbar} R_1^{(\alpha)} R_2^{(\alpha)} \left(\frac{E_c}{k_B T}\right)^2, \tag{4.107}$$

which expresses the inelastic cotunneling [55, 59]. We emphasize that the resistance $R_\alpha$ for sequential tunneling is proportional to the *sum* of the resistances, $R_1^{(\alpha)} + R_2^{(\alpha)}$, of the two junctions, while $R_\alpha$ in the cotunneling case is proportional to the *product* of the resistances, $R_1^{(\alpha)} R_2^{(\alpha)}$, of the two junctions.

The TMR ratio $\Delta R/R_F$, where $\Delta R = R_A - R_F$ is the difference in the resistances of the A and F alignments, is given by

$$\frac{\Delta R}{R_F} = r_1^{(F)}\left[\frac{2P_1 P_2}{1 - P_1 P_2}\right] + \left(1 - r_1^{(F)}\right)\left[\frac{2P_2 P_3}{1 - P_2 P_3}\right], \qquad (k_B T \gg E_c), \tag{4.108}$$

in the sequential-tunneling regime, where $r_1^{(F)}$ is the reduced resistance of the 1st junction in the F alignment defined by

$$r_1^{(F)} = R_1^{(F)}/[R_1^{(F)} + R_2^{(F)}], \qquad (0 < r_2^{(F)} < 1).$$

The TMR for the sequential tunneling is given by the average of the TMRs at the two junctions weighted by the reduced resistances. In the cotunneling regime, on the other hand, we have

$$\frac{\Delta R}{R_F} = \frac{2\left(P_1 P_2 + P_2 P_3\right)}{(1 - P_2 P_3)(1 - P_2 P_3)}, \qquad (k_B T \ll E_c). \tag{4.109}$$

In the case of symmetric junction, $R_1^{(\alpha)} = R_2^{(\alpha)} = R_T^{(\alpha)}$ ($r_1^{(\alpha)} = 1/2$) and $P_1 = P_2 = P$, the TMR ratio is given by

$$\frac{\Delta R}{R_F} = \frac{2P^2}{1 - P^2}, \qquad (k_B T \gg E_c) \tag{4.110}$$

in the sequential-tunneling regime, while in the cotunneling regime

$$\frac{\Delta R}{R_F} = \frac{4P^2}{(1 - P^2)^2}, \qquad (k_B T \ll E_c). \tag{4.111}$$

Therefore, the TMR for $k_B T \ll E_c$ is enhanced by $2/(1 - P^2)$ compared with TMR for $k_B T \gg E_c$ in the absence of the CB effect. For example, if $Co_{50}Fe_{50}$

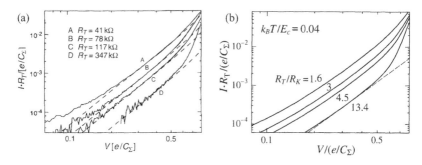

Figure 4.22: (a) Current-voltage ($I$-$V$) characteristic measured in non-magnetic SET by Geerligs *et al.* [56]. (b) Calculated $I$-$V$ curves for comparison with the experiments. The dashed line indicates $I \propto V^3$, representing the co-tunneling contribution. The deviation of the solid curves from the dashed line at low bias is due to the thermally assisted tunneling contribution.

is used for the electrodes and $P = 0.51$ for the spin polarization in Table 4.1 (also see Fig. 3 in Sec. 3.1), an enhancement of $\Delta R/R_F$ from 70% to 190% is expected by the CB effect.

Before going into discussions on TMR, we briefly mention a non-magnetic SET device. If the spin-dependent tunnel resistance $R_i^{(\alpha)}$ is replaced with the non-magnetic one $R_T$, the present formulation is applicable to non-magnetic SET devices. To demonstrate this, we calculate the current-voltage ($I$-$V$) characteristic for different values of $R_T$ as shown in Fig. 4.22(b). Our calculation reproduces very well the experimental results on a non-magnetic SET made of aluminum by Geerligs *et al.* [56] as shown in Fig. 4.22(a).

Let us now discuss the temperature, gate voltage, and bias voltage dependencies of the resistance and TMR in the magnetic SET. Figure 4.23(a) shows the temperature dependence of the resistance $R_\alpha$ at $V = V_g = 0$ for three values of tunnel resistance $R_T^{(\alpha)}/R_K$, where $R_K = h/e^2 \simeq 25.8\,\mathrm{k\Omega}$. The curves clearly show the crossover around $k_B T/E_c = 0.1$, below which the cotunneling gives the dominant contribution since the sequential tunneling is exponentially suppressed and the resistance shows the $1/T^2$ dependence, and above which sequential tunneling is recovered. Note that, in the limit of $R_T^{(\alpha)}/R_K \to \infty$, cotunneling disappears and $R_\alpha$ follows the classical expression $R_\alpha/2R_T^{(\alpha)} \sim (k_B T/E_c)\sinh(E_c/k_B T)$ in the temperature range $0.1 < k_B T/E_c < 0.4E_c$, above which $R_\alpha$ follows the well-known expression $R_\alpha/2R_T^{(\alpha)} \sim 1 + (E_c/3k_B T) + (2/45)(E_c/k_B T)^2 + \cdots$. The thin dashed line indicates the resistance at the gate voltage of $eV_g/E_c = C_\Sigma/C_g$, where the charging energy is canceled out by this gate voltage and the CB is neutralized. Figure 4.23(b) shows the TMR ratio as a function of temperature for different values of $R_T^{(A)}$ and $R_T^{(F)}$ keeping their ratio $R_T^{(A)}/R_T^{(F)} = 2$ ($P = 0.58$).

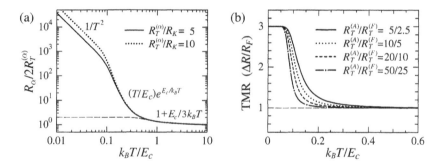

Figure 4.23: (a) Resistance and (b) tunnel magnetoresistance as functions of temperature at $V = 0$ and $V_g = 0$, where $\alpha$ is $F$ (ferromagnetic) or $A$ (antiferromagnetic) alignment of magnetization in the electrodes. The thin dashed lines indicate the $R_\alpha$ and TMR on resonance.

As temperature is lowered, the TMR is enhanced from $\Delta R/R_F = 1$ where the sequential tunneling is dominant, to $\Delta R/R_F = 3$ where the cotunneling is dominant. In the crossover region both tunneling processes contribute to the tunnel current.

Figure 4.24(a) shows the resistance $R_\alpha$ as a function of gate voltage at $V = 0$ and for selected values of temperature. Around $eV_g/E_c = 0$ and $\pm 10$ the resistance $R_\alpha$ increases rapidly with decreasing temperature. On the other hand, the values of the minima in $R_\alpha$ at $eV_g = (C_\Sigma/C_g)E_c = \pm 5E_c$ are almost unchanged. The minima in the resistance correspond to well-known conductance resonance peaks, where the neighboring charge states, $n = 0$ and $n \pm 1$, have the same electrostatic energy, and there is no charging effect. Figure 4.24(b) shows the TMR ratio as a function of gate voltage. We see that the

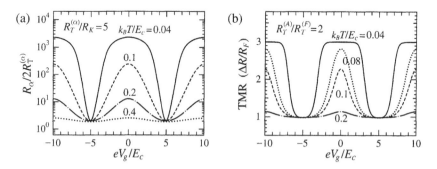

Figure 4.24: (a) Resistance and (b) tunnel magnetoresistance as functions of gate voltage $V_g$ at zero bias voltage $V = 0$, where $\alpha$ is $F$ (ferromagnetic) or $A$ (antiferromagnetic) alignment of magnetization in the electrodes.

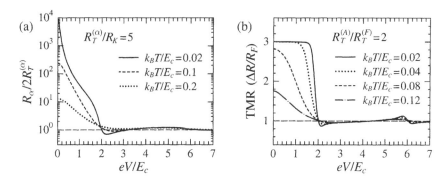

Figure 4.25: (a) Resistance and (b) tunnel magnetoresistance as functions of bias voltage $V$ at zero gate voltage $V_g = 0$, where $\alpha$ is $F$ (ferromagnetic) or $A$ (antiferromagnetic) alignment of magnetization in the electrodes.

behavior of the TMR curves is well correlated with that of $\log(R_\alpha)$. In the CB regime ($k_B T < 0.2 E_c$) the TMR around $V_g = 0$ progressively increases from $\Delta R/R_F = 1$ to $\Delta R/R_F = 3$ as temperature is lowered, while the TMR around $eV_g/E_c = \pm 5$ is unchanged. This indicates that the TMR is continuously controlled by the gate voltage.

Figure 4.25(a) shows resistance $R_\alpha$ as a function of bias voltage at $V_g = 0$ and for three different temperatures. At low temperatures, the resistance increases rapidly with decreasing voltage in the CB regime ($eV/E_c < 2$), and shows a power law dependence, $R_\alpha \propto 1/V^2$, as expected for cotunneling process. Figure 4.25(b) shows the TMR ($\Delta R/R_F$) as a function of bias voltage. As temperature is decreased, the TMR is enhanced from $\Delta R/R_F = 1$ to $\Delta R/R_F = 3$. At $k_B T/E_c = 0.02$, the enhanced TMR is constant up to $eV/E_c \sim 1.6$, whereas the corresponding resistance is reduced by several orders of magnitude.

### 4.3.3.3   Asymmetric double tunnel junction

Asymmetry in double tunnel junctions has a significant effect on the temperature and bias dependence of the TMR. In an asymmetric tunnel junction, the height and thickness of the tunnel barriers and/or the relative density of states of majority and minority spin electrons of the ferromagnets are different, which are characterized by the different values of the tunnel resistance and those of the spin-polarization in the junction. The effect of the asymmetry on the TMR is different in the sequential and in cotunneling regime. In the sequential regime ($k_B T \gg E_c$), the TMR strongly depends on the relative values of the tunnel resistance at junctions 1 and 2, because from Eq. (4.108) the TMR is given by the sum of the TMR at each tunnel junction weighted by the relative resistances $r_i^{(F)}$ of the $i$th junction, indicating that the TMR has a

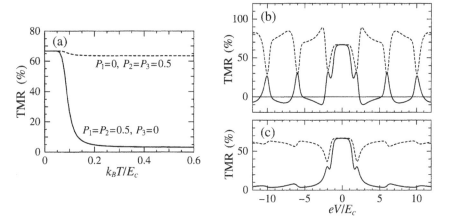

Figure 4.26: (a) Temperature dependence of tunnel magnetoresistance (TMR) at zero bias and gate voltages ($V = V_g = 0$). (b) Bias dependence of TMR at $k_B T / E_c = 0.05$ and $V_g = 0$: In the upper panel the TMR is defined by $(G_A - G_F)/G_A$, and in the lower panel the TMR is $(I_F - I_A)/I_A$. The solid (dashed) curve represents the TMR of the junction with the right (left) hand electrode being a normal metal, in which the resistance of the second junction is much larger than that of the first junction, i.e., $R_2^{(F)}/R_1^{(F)} = 20$.

large contribution from the tunnel junction with a larger junction resistance. Therefore the magnitude of the TMR in the sequential regime is very much influenced by the difference in the junction resistances, especially when the spin polarizations $P_1$ and $P_3$ of the outer electrodes are different. In contrast, in the cotunneling regime ($k_B T \ll E_c$), the TMR has no explicit dependence on the relative magnitude of the tunnel resistance as seen in Eq. (4.109).

The most dramatic difference in the TMR between the sequential tunneling and cotunneling regimes arises when one of the left and right electrodes is a normal metal ($P_1 = 0$ or $P_3 = 0$) and others are ferromagnets, and that one of the junction resistance is much larger than the other. Figure 4.26(a) shows the temperature dependence of the TMR for the case of $P_3 = 0$, $P_1 = P_2 = 0.5$ and $R_2/R_1^{(F)} = 20$ by the solid curve. In this case, the TMR in the sequential regime is given by $\Delta R / R_F = [2P_1 P_2 / (1 - P_1 P_2)] r_1^{(F)}$ with a small value of $r_1^{(F)} = 1/21$, while in the cotunneling regime $\Delta R / R_F = 2P_1 P_2 / (1 - P_1 P_2)$. Therefore the TMR has a large enhancement from $\Delta R / R_F = 3.2\%$ where the sequential tunneling is dominant, to $\Delta R / R_F = 67\%$ where the cotunneling is dominant. In Fig. 4.26(b), the bias dependence of the TMR is shown by the solid curve. In the sequential regime, a strong oscillation of the TMR appears due to the asymmetry in the tunnel junction, as pointed out by Barnas and Fert [40]. The oscillation becomes stronger (weaker) as the junction asymmetry is increased (decreased). In a symmetric junction, the oscillation is very small but still visible as seen in Fig. 4.25(b).

If one interchanges the left and right electrodes, i.e., the left electrode is chosen to be a normal metal ($P_1 = 0$), the behavior of the TMR is drastically changed as shown by the dashed curves in Fig. 4.26(a); the TMR has little temperature dependence, since the TMR is regulated by the second junction and given by $\Delta R/R_F = [2P_2P_3/(1 - P_2P_3)]\, r_2^{(F)}$, which is slightly smaller than the TMR in the cotunneling regime by the factor $r_2^{(F)} = 20/21$. The corresponding bias dependence of the TMR is shown in Figs. 4.26(b) and (c) by the dashed curves. In the cotunneling regime, the dashed curve coincides with the solid curve, as expected. In the sequential regime, however, the oscillations are dramatically different between the two curves. It should be noted that the resistance ratios satisfy the inequality $R_1^{(A)}/R_2^{(A)} < R_1^{(F)}/R_2^{(F)}$ in the case of $P_1 = 0$ and $R_1^{(A)} = R_1^{(F)} = R_1$, while $R_1^{(A)}/R_2^{(A)} > R_1^{(F)}/R_2^{(F)}$ in the case of $P_2 = 0$ and $R_2^{(A)} = R_2^{(F)} = R_2$. This difference leads to the oscillations in a reversed way as seen in the solid and dashed curves in Fig. 4.26(b).

### 4.3.3.4   Discussions

There have been several experimental evidences for the enhancement of the TMR in the small ferromagnetic double tunnel junctions. Ono *et al.* have measured the TMR in the Ni/NiO/Co/NiO/Co double junctions and observed Coulomb oscillations in the $R_\alpha$ vs $V_g$ curves [60, 67]. Off resonance (i.e., at the peaks of $R_\alpha$ in Fig. 4.24) they found that the TMR ratio, $\Delta R/R_F = (R_A - R_F)/R_F$, is enhanced to 40%, which is larger than the value of 17.5% expected from $P_{Co} = 0.35$ and $P_{Ni} = 0.23$ in the absence of the CB effect, as shown in Fig. 4.27. The present theory explains this enhancement since $\Delta R/R_F = 4P_{Co}P_{Ni}/(1 - P_{Co}P_{Ni})^2 = 0.38$, i.e., the TMR is 38%. However,

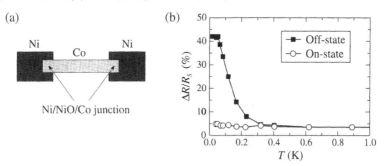

Figure 4.27: (a) Schematic view of a magnetic SET junction consisting of Ni electrodes and Co island with the junction area $\sim 200\,\mathrm{nm}^2$. (b) Tunnel magnetoresistance (TMR) as a function of temperature measured by Ono *et al.* in Ni/NiO/Co/NiO/Ni [60, 67].

they found the TMR of ~4% at resonance (i.e., at the bottoms of in Fig. 4.24), which is considerably smaller than the expected TMR of 17.5%. A reduction in TMR at resonance may occur in case of strong tunneling $R_T^{(\alpha)} < R_K$. Recent experiments and theories for the SET have shown that the conductance in the strong tunneling case significantly deviates from that in the weak tunneling case $R_T^{(\alpha)} \gg R_K$. According to the theory of König *et al.* [58], on resonance the bare value of the parameters, $E_c$ and $R_T^{(\alpha)}$, are renormalized to make the conductance to be logarithmic temperature dependence, yielding the TMR at low temperatures

$$\left(\frac{\Delta R}{R_F}\right)_{\text{on}} = \frac{2P^2}{(1 - P^2)} \left\{ 1 - \frac{R_K}{2\pi^2 R_T^{(F)}} \left[ \gamma_E + \ln\left(\frac{E_c}{\pi T}\right) \right] \right\}, \qquad (4.112)$$

with $\gamma_E$ Euler's constant. Therefore the value of TMR at low temperatures ($k_B T \ll E_c$) is considerably reduced from that at high temperatures. If this is the case for the Ni/NiO/Co/NiO/Co double junctions, we can explain the small value of TMR mentioned above. In contrast, the TMR off resonance remains unaltered [68].

Recently, Wang and Brataas [69] have studied transport through a ferromagnetic SET and represented the resistance as a path integral in order to investigate the strong tunneling regime beyond the low-order sequential tunneling and cotunneling regimes. Using Monte Carlo simulations, they found a large MR ratio at sufficiently low temperatures.

### 4.3.4 Magnetic granular systems with Coulomb blockade

Magnetic granular materials containing ferromagnetic granules of nanometer size in an insulating matrix are unique systems, since the magnetic moments of the granules are randomly oriented, and the charging energy $E_c$ of the granules is broadly distributed due to the variation of granule size $d$ through the relation $E_c \sim e^2/\epsilon_{\text{eff}}d$, where $\epsilon_{\text{eff}}$ is the effective dielectric constant. The conduction of charge carriers in the granular structures involves a large number of Co granules with different size. Once a charge carrier enters a large granule, which is surrounded by smaller ones, the carrier cannot tunnel to the neighboring small granules due to the CB at low temperatures. However, because of the distribution of granule size, the carrier can find a large granule of nearly equal size in the neighborhood, which is separated by a number of small granules. Then, it is energetically allowed to transfer the carrier through the intervening small granules to the large granule using the higher-order tunneling process, where several successive tunneling processes of single electrons are involved in the intermediate state. The successive tunneling is much easier in an aligned magnetic moments in a high magnetic field than in a randomly oriented ones in zero field. Therefore, we expect a large enhancement of MR due to the higher-order tunneling process. In the granular systems, the existence of higher-order tunneling processes and the enhancement of MR is a consequence of their granular structure with broad distribution of granule size. In the following, we preset a model for magnetic granular systems which provides a consistent explanation for the temperature and bias-voltage dependence of the resistivity and the MR observed in Co-based granular films.

### 4.3.4.1   Model for the granular systems

The characteristic temperature dependence of the electrical resistivity, $\ln \rho_0(T)$ $\propto 1/\sqrt{T}$, has been observed in magnetic as well as non-magnetic granular materials in a wide range of metal-nonmetal compositions below the percolation threshold. Sheng *et al.* [70] have derived the temperature dependence based on the model that granules on each conduction path are equal in size $d$ and separated by barrier thickness $s$, keeping their ratio $s/d$ (or equivalently $E_c s$) constant for a given composition. In this model, the electrical conductivity $\sigma_0(T)$ at high temperatures is contributed from tunneling between small granules, while at low temperatures $\sigma_0(T)$ is contributed from tunneling between large granules. An extension to the magnetic granular systems has been made by Inoue and Maekawa [51], who incorporate the effect of spin-dependent tunneling into the model and obtain the spin-dependent conductivity $\sigma_m(T) = (1 + P^2 m^2)\, \sigma_0(T)$ and the MR ratio

$$\Delta\rho/\rho_0 = P^2 m^2 / \left(1 + P^2 m^2\right), \tag{4.113}$$

where $P$ is the spin-polarization of the granules and $m = M/M_s$ the magnetization normalized by the saturation magnetization $M_s$. The $m = 0$ ($m = 1$) corresponds to the randomly (fully) oriented magnetic moments of granules in zero (high) magnetic field. The formula (4.113) reproduces well the behavior of $\Delta\rho/\rho_0 \propto m^2$ observed in magnetic granular systems [71].

Recently, Mitani *et al.* [50] have found that the MR in Co-based granular films exhibits a strong temperature dependence at low temperatures and is enhanced more than 20% at 4.2 K, which is twice as large as the value of 11% estimated from Eq. (4.113) using $P = 0.34$ for Co [3]. It should be noted that the above model makes a gross simplification for the actual conduction paths by neglecting tunneling between granules of unequal size. We extend the model to include tunneling between those granules, which plays a crucial role for the MR in the magnetic granular systems.

In the granular systems with a broad distribution of granule size, large granules are distant from each other due to their low number density, i.e., the larger the granule size is, the more separated the granules are, so that there is a number of smaller granules between large granules as shown in Fig. 4.28(a). For modeling the structural feature of granular systems, we assume that, when

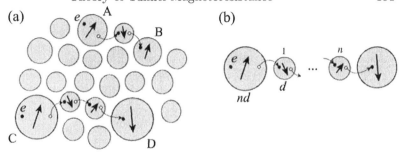

Figure 4.28: (a) Schematic diagram of a granular structure and a higher-order tunneling process whereby a charge carrier is transferred from the charged large granule A (C), via the small ones, to the neutral large one B (D), leaving behind an electron and a hole excitation on the granules. (b) Conduction path used for the calculation of the electrical conductivity.

there are large granules with size $n\bar{d}$ and charging energy $\bar{E}_c/n$, they are separated by the number of $n$ intervening granules with the average size $\bar{d}$ and charging energy $\bar{E}_c$, as shown in Fig. 4.28(b). The tunneling distance between the neighboring granules is $s = 2n\bar{s}/(n+1)$, where $\bar{s}$ is a mean separation if all the granules would have the average size $\bar{d}$.

(a) **Temperature dependence of MR**    We calculate the temperature dependence of the conductivity $\sigma_m(T)$ at zero bias voltage $(V = 0)$ in our model structure of granular systems. The tunnel current in the zero bias limit is dominated by thermally activated charge carriers. In the conduction path in Fig. 4.28(b), the carriers occupy mostly the large granule of charging energy $\bar{E}_c/n$ in the probability proportional to the Boltzmann factor $\exp\left(-\bar{E}_c/2nk_BT\right)$. Since the charged large granules are separated by the smaller granules, the ordinary tunneling of an electron from the large granule to the small granule causes an increase of charging energy and thus suppressed by the Coulomb blockade at low temperatures. In this regime, the dominant contribution to $\sigma_m(T)$ comes from higher-order tunneling processes where the carrier is transferred to the neighboring large granule through the intervening small ones using the successive tunneling of single electrons, i.e., the cotunneling of $(n+1)$ electrons, leaving behind an electron and a hole excitation in each granule (see Fig. 4.28(b)).

In the higher-order process, the total tunneling probability between the large granules is given by the product of those between the neighboring granules:

$$\left\langle \prod_{i=0}^{n} e^{-2\kappa s} \left(1 + P^2 \cos\theta_{i,i+1}\right) \right\rangle, \tag{4.114}$$

where $\kappa$ is a tunneling parameter related to the barrier height, $\theta_{i,i+1}$ is the angle between the magnetic moments $\boldsymbol{\mu}_i$ and $\boldsymbol{\mu}_j$ of the neighboring granules

$i$ and $i + 1$, and $\langle \cdots \rangle$ denotes the average over the direction of the magnetic moments. If Eq. (4.114) is expanded with respect to $P^2$ up to the first order and the relation $\langle \cos \theta_{i,i+1} \rangle = \langle \boldsymbol{\mu}_i \cdot \boldsymbol{\mu}_j \rangle / \mu_i \mu_j = m^2$ is used, Eq. (4.114) is well approximated by the form $\left[ (1 + P^2 m^2) \, e^{-2\kappa s} \right]^{n+1}$.

Summing up all of the higher-order processes of the spin-dependent tunneling, we obtain the zero bias conductivity

$$\sigma_m(T) \; \propto \; \sum_n e^{-\bar{E}_c/2nk_BT} \left[ (1 + P^2 m^2) \, e^{-2\kappa s} \right]^{n+1} g_n(T), \qquad (4.115)$$

where $g_n(T) \propto (\pi k_B T)^{2n}$ is a temperature-dependent factor arising from the electron and hole excitations available in the energy interval of $\pi T$ around the Fermi level [50]. In the following, we neglect the temperature dependence of $g_n(T)$, because $g_n(T)$ is spin-independent and its temperature dependence is much weaker than the Boltzmann factor at low temperatures. In Eq. (4.115), the exponential factor $\exp[4\kappa n \bar{s} - \bar{E}_c / 2nk_B T]$, where we introduce the mean separation $\bar{s} = (n + 1)s/2n$ of granules with radius $\bar{d}$, has a sharp peak at $n = n^*$:

$$n^* = \left( \bar{E}_c / 8\kappa \bar{s} k_B T \right)^{1/2}. \qquad (4.116)$$

At low temperatures ($k_B T \ll E_c$), $n^*$ becomes so large that $n$ is treated as a continuous valuable. Replacing the summation by the integration and using the method of steepest descent in Eq. (4.115), we obtain the spin-dependent conductivity

$$\sigma_m(T) \propto \left( 1 + P^2 m^2 \right)^{n^*+1} \exp \left[ -2 \left( 2\kappa \bar{s} \bar{E}_c / k_B T \right)^{1/2} \right]. \qquad (4.117)$$

Figure 4.29(a) shows the resistivity $\rho_0(T)$ in zero field ($m = 0$). In the calculation, we take the values of the parameters: $2\kappa \bar{s} = 3.5$, $g = 0.3$, and $\bar{E}_c = 100 \, \text{K}$. Our model reproduces the resistivity of $Co_{36}Al_{22}O_{42}$ granular films [50].

Due to the higher-order processes, the spin-dependent part of $\sigma_m(T)$ in Eq. (4.117) is amplified to the $(n^* + 1)$th power of $(1 + P^2 m^2)$, so that the resistivity $\sigma_m(T)$ is sensitive to the change of $m$ by application of a magnetic field. Substituting Eq. (4.117) into the MR ratio, $\Delta \rho / \rho_0 = 1 - \sigma_m(T) / \sigma_0(T)$, we have the MR ratio in the granular systems

$$\Delta \rho / \rho_0 = 1 - \left( 1 + m^2 P^2 \right)^{-(n^*+1)}. \qquad (4.118)$$

The MR ratio has rather a simple form because of the cancellation of $\sigma_0(T)$, which is the complicated function of temperature and granular structure, is dropped out by taking the ratio.

Figure 4.29(b) shows the MR ratio for $m = 1$ and $P = 0.31$. The value of $P$ is slightly smaller than that of $Co = 34\%$ measured by tunnel spectroscopy [3]. Other parameters are the same as those in Fig. 4.29(a). The solid dots represent the experimental data of $Co_{36}Al_{22}O_{42}$ by Mitani *et al.* The dashed line

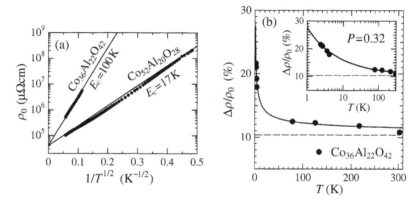

Figure 4.29: (a) Resistivity and (b) magnetoresistance as functions of temperature. The solid circles indicate the experimental results measured by Mitani *et al.* [50].

indicates the value of the MR ratio in Eq. (4.113). For small $P^2$, Eq. (4.118) is approximated to be a convenient formula

$$\Delta\rho/\rho_0 \approx P^2 m^2 \left(1 + \sqrt{C/T}\right), \qquad (4.119)$$

with $C(= \bar{E}_c/8\kappa s k_B)$ being a constant. The dashed curves in Fig. 4.29 indicate the values calculated from the formula (4.119) for $m = 1$, $P = 0.325$, and $C = 2.6\,\text{K}$. The anomalous increase of $\Delta\rho/\rho_0 \sim 1/\sqrt{T}$ at low temperatures is due to the onset of higher-order processes between larger granules, i.e., $n^* \propto 1/\sqrt{T}$. As seen in the curve, the MR grows rapidly around $\sim 10\,\text{K}$ well below $E_c = 95\,\text{K}$. Similar behavior is seen in a ferromagnetic SET [65]. At high temperatures ($k_B T \gg \bar{E}_c$), Eq. (4.118) reproduces the formula (4.113) in the absence of the charging effect.

(b) **Bias voltage dependence of MR** We next discuss the bias voltage ($V$) dependence of the conductivity $\sigma_m(V)$ in the CB regime. When a finite voltage $V$ is applied to the granular system, the voltage drop $\Delta V$ between the large granules in the model system of Fig. 4.28(b) is given by $\Delta V = (2n/N_g)V$, where $N_g$ is the average number of granules along a conduction path. If one uses the result of Averin and Odintsov [35], who first discussed the higher-order tunneling in a 1D array of islands (granules), then one can extract the bias-dependent part of the current through the array of granules in Fig. 4.28(b). The result is incorporated into the current expression as

$$I_m(V) \propto \sum_n e^{-\bar{E}_c/2n k_B T} \left[(1 + P^2 m^2) e^{-2\kappa s}\right]^{n+1}$$

$$\times \; e\Delta V \prod_{\nu=1}^{n} \left[1 + \left(\frac{e\Delta V}{2\pi\nu k_B T}\right)^2\right]. \qquad (4.120)$$

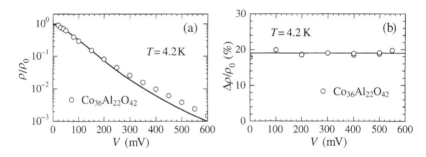

Figure 4.30: Bias voltage dependence of (a) resistivity $\rho_0(V)$ and (b) MR ratio $\Delta\rho/\rho_0$. The parameters are the same as those in Fig. 4.29. The open circles indicate the experimental results of $Co_{36}Al_{22}O_{42}$ film at 4.2 K [50].

By interpolating the conductance $\sigma_m(V) = dI_m/dV$ between the limiting cases of $e\Delta V/k_BT \to 0$ and $e\Delta V/k_BT \gg 1$, we obtain

$$\sigma_m(V) \propto \sum_n e^{-\frac{E_c}{2nk_BT}} \left[(1 + P^2m^2)\, e^{-2\kappa s}\right]^{n+1} \left[1 + \left(\frac{2eV}{N_g\pi k_BT}\right)^2\right]^n . \quad (4.121)$$

Following the same procedure as in deriving $\sigma_m(T)$ in Eq. (4.117), we obtain the bias-dependence of the conductivity

$$\sigma_m(V) = \sigma_m(T) \left[1 + (2eV/N_g\pi k_BT)^2\right]^{n^*} . \quad (4.122)$$

The $\sigma_m(V)$ exhibits a power law dependence $(1/V)^{2n^*}$ for $k_BT < eV/N_g$. Note that the conductance of SET behaves as $1/V^2$ in the CB regime [55, 59].

Figure 4.30(a) shows the resistivity $\rho_0(V) = 1/\sigma_0(V)$ calculated from Eq. (4.122) for $T = 4.2$ K and $N_g = 160$. The value of $N_g = 160$ is somewhat smaller than the value $200 \sim 300$ estimated from the image of TEM micrographs [72]. The calculated resistivity decreases steeply with increasing $V$, in agreement with the experimental data. A deviation of the calculated values from the experimental ones starts at about 300 mV and become enlarged for higher voltages, suggesting that the density of charge carriers created by the bias voltage are comparable to or larger than that of thermally activated ones at higher voltages.

The $V$-dependence of the MR ratio is also given by Eq. (4.118). Figure 4.30(b) shows the $V$-dependence of the MR ratio. The enhanced MR ratio is maintained up to higher voltages, in agreement with the experimental data of $Co_{36}Al_{22}O_{42}$. The constant MR may originate from the large number of granules along the conduction paths in the granular films, in which the voltage drop between neighboring granules $\Delta V \sim V/N_g$ is small for large number of $N_g$. The voltage drop $\Delta V$ at the bias voltage $V = 500$ mV is about 3 mV for $N_g = 160$ and its corresponding temperature is $\sim 40$ K, which is smaller than the charging energy $\bar{E}_c \sim 100$ K. We note that the enhanced MR is nearly constant up to 500 meV, whereas the corresponding resistance is reduced by several orders of magnitude, This is in contrast with ferromagnetic tunnel junctions with macroscopic size, where both MR and resistance decrease gradually with increasing bias-voltage [7].

#### 4.3.4.2 Tunnel junction with granules in the barriers

There has been a number of experiments on ferromagnetic tunnel junctions in which ferromagnetic granules such as Co are dispersed in the insulating layer of the tunnel barrier [62, 63, 64]. In these hybrid tunnel junctions, thermodynamical fluctuations in the magnetic moments of granules, i.e., superparamagnetism, is seen in the higher-order process of the spin-dependent tunneling. To demonstrate this, we consider a tunnel junction containing a small granule in the barrier as shown in Fig. 4.31. The magnetic moments of the electrodes takes either ferromagnetic (F) or antiferromagnetic (A) alignment, depending on the strength of magnetic field. In zero field, the magnetic moments $\mu_g$ of the granules are in a superparamagnetic state, i.e., thermally rotating, or randomly oriented. In the CB regime $k_B T \ll E_c$, the current is dominated by the cotunneling through the granules. Using the spin-dependent tunneling probability in Eq. (4.114) with $n = 1$, the conductance $G_F(H)$ in the F alignment is written as

$$G_F(H) = G_0(T) \left[ 1 + 2P^2 \langle \cos \theta \rangle + P^4 \langle \cos^2 \theta \rangle \right], \qquad (4.123)$$

while in the A alignment

$$G_A(H) = G_0(T) \left[ 1 - P^4 \langle \cos^2 \theta \rangle \right], \qquad (4.124)$$

where $\theta$ is the angle between the moment of the left electrode and the granule and $\langle \cdots \rangle$ denotes the spatial and thermal averages. The $\langle \cos \theta \rangle$ represents the polarization of the moment $\mu_g$ in the direction of $H$ and is given by the Langevine function $L(x)$:

$$\langle \cos \theta \rangle = L(x) = \coth(x) - 1/x, \quad x = \mu_g H / k_B T. \qquad (4.125)$$

The term $\langle \cos^2 \theta \rangle$ appears as a consequence of the cotunneling process and is given by

$$\langle \cos^2 \theta \rangle = 1 - (2/x)L(x), \qquad (4.126)$$

which represents the correlation between the moments of granules and the electrodes. This enables us to extract the degree of thermal or spatial fluctuation of the moments $\mu_g$ by measuring the conductance (or resistance) in the CB regime.

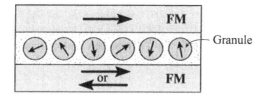

Figure 4.31: Tunnel junction containing small granules in the insulating barrier. The magnetic moments of the granule are in a superparamagnetic state, i.e., thermally rotating, or randomly oriented, and those of the top and bottom electrodes take either ferromagnetic (F) or antiferromagnetic (A) alignment.

In zero field $(H = 0)$, the magnetic moments are randomly oriented, where $\langle \cos \theta \rangle = 0$ and $\langle \cos^2 \theta \rangle = 1/3$, so that

$$G_F(H) = G_0(T) \left[1 + P^4/3\right],\tag{4.127}$$

in the F alignment, while

$$G_A(H) = G_0(T) \left[1 - P^4/3\right],\tag{4.128}$$

in the A alignment. Therefore the cotunneling gives rise to the difference in the conductance at $H = 0$, and the resistance ratio is given by

$$(R_A/R_F)_{H=0} = 1 + (2/3)P^4.\tag{4.129}$$

Figure 4.32(a) shows the resistances, $R_F(H)$ and $R_A(H)$, as a function of magnetic field $H$ for the spin-polarization $P = 50\%$ in the electrodes and the granule. Figure 4.32(b) shows the $H$-dependence of the resistance obtained from the curves in Fig. 4.32(a), when the magnetic moments of the electrodes

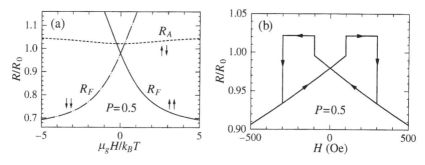

Figure 4.32: (a) Resistance $R_F(H)$ and $R_A(H)$ as a function of magnetic field $H$. (b) Resistance curve obtained from the curves in (a) when the magnetic moments of the electrodes are parallel for fields $|H|$ between 100 Oe and 300 Oe and antiparallel otherwise.

are in the A alignment for fields $|H|$ between 100 Oe and 300 Oe and otherwise in the F alignment. Here, we assume $\mu_g = 1000\mu_B$ for the magnetic moment of the granule and take $T = 67$ K for temperature, yielding a scale of magnetic field $k_B T/\mu_g = 1$ kOe for a thermally rotating granule. The resistance curve in Fig. 4.32(b) is similar to that of recent experiments in hybrid junctions [63].

## 4.4 Spin accumulation

### 4.4.1 Spin injection into a normal metal

Spin injection into a normal metal without tunneling barriers was predicted by Aronov [73] and observed by Johnson and Silsbee [74, 75]. The spin injection technique was developed using a three terminal device called the bipolar spin transistor (see Fig. 4.33(b)). This device is based on the non-equilibrium effect of spin accumulation generated by a spin-polarized current from a ferromagnet (FM1) into a non-magnetic metal (N) such as Al and Au. In the spin injection experiments, the spin-injection signals were detected with another ferromagnet (FM2), which is deposited on the other side of N, by changing the relative orientation of magnetizations in FM1 and FM2. The spin diffusion length $\lambda_S$ of non-magnetic metals was obtained by varying the thickness of the metal between the ferromagnets. In the following, we consider a TMR device as shown in Fig. 4.33(a), which is intimately related to the CPP GMR in magnetic multilayers [76]. The TMR device is apparently different from the original bipolar spin transistor because spins are injected into or extracted from the normal metal at the left and right junctions. However, these two devices are essentially the same from the view point of the Onsager relations [77].

Let us examine the spin injection and accumulation in a double tunnel junction (FM1/N/FM2) shown in Fig. 4.33(a). The electrodes of FM1 and FM2 are made of different ferromagnets and the central one of N is a non-magnetic metal with thickness $d$. The magnetization of the left FM1 is chosen to point up and that of the right FM2 is either up (parallel) or down (antiparallel). The bias injection current $I_{\text{inj}}$ flows through the junction.

When there exists a gradient in the electrochemical potential $\mu_\sigma(x)$ for electrons with spin $\sigma$ in N, the electric current $j_\sigma$ of spin $\sigma$ flows according to

$$j_\sigma(x) = -\frac{\sigma_N}{2e} \frac{\partial}{\partial x} \mu_\sigma(x), \qquad (4.130)$$

where $\sigma_N$ is the conductivity of N. If there is a spin-flip scattering in N, the divergence of the current in the spin $\sigma$ channel is balanced with the difference between the spin relaxation rate from spin $\sigma$ to spin $-\sigma$ and that from spin $-\sigma$ to spin $\sigma$ in the steady state

$$\frac{\partial}{\partial x} j_\sigma(x) = -\frac{e}{2\tau_S} \left[ n_\sigma(x) - n_{-\sigma}(x) \right], \qquad (4.131)$$

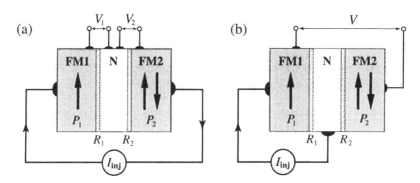

Figure 4.33: Schematic drawing of spin injection devices of ferromagnet/normal-metal/ferromagnet junctions: (a) TMR device and (b) Spin-bipolar device.

where $n_\sigma$ is the local spin density with spin $\sigma$ in N and $\tau_S$ is the spin relaxation time. Taking the difference of Eq. (4.131) with spin-up and spin-down, we have

$$\frac{\partial}{\partial x}\left[j_\uparrow(x) - j_\downarrow(x)\right] = -\frac{e\mathcal{D}_N}{\tau_S}\left[\mu_\uparrow(x) - \mu_\downarrow(x)\right], \qquad (4.132)$$

where $\mathcal{D}_N$ is the density of states for electrons per spin in N. Taking the sum of Eq. (4.131) with spin-up and spin-down, we have the current conservation

$$\frac{\partial}{\partial x}\left[j_\uparrow(x) + j_\downarrow(x)\right] = 0. \qquad (4.133)$$

We introduce the splitting $\delta\mu(x)$ and average $\bar{\mu}(x)$ of the chemical potentials

$$\delta\mu(x) = \frac{1}{2}\left[\mu_\uparrow(x) - \mu_\downarrow(x)\right], \qquad \bar{\mu}(x) = \frac{1}{2}\left[\mu_\uparrow(x) + \mu_\downarrow(x)\right]. \qquad (4.134)$$

The spin density $\mathcal{S}(x)$ induced by the splitting of the chemical potential $\delta\mu(x)$ and the spin current $j_{\text{spin}}(x) = j_\uparrow(x) - j_\downarrow(x)$ flowing in N are given by

$$\mathcal{S}(x) = \mathcal{D}_N\delta\mu(x), \qquad j_{\text{spin}}(x) = -\frac{\sigma_N}{e\mathcal{D}_N}\frac{\partial}{\partial x}\mathcal{S}(x). \qquad (4.135)$$

From Eqs. (4.130) and (4.132), we obtain the equations for $\delta\mu(x)$ and $\bar{\mu}(x)$ :

$$\frac{\partial^2}{\partial x^2}\delta\mu(x) = \frac{\delta\mu(x)}{\lambda_S^2}, \qquad \frac{\partial^2}{\partial x^2}\bar{\mu}(x) = 0, \qquad (4.136)$$

where $\lambda_S = \sqrt{D\tau_S}$ is the spin-diffusion length with $D$ the diffusion constant satisfying the Einstein relation $D = \sigma_N/[2e^2\mathcal{D}_N]$ in the normal metal. The second equation in Eq. (4.136) has the solution $\bar{\mu}(x) = (ej/\sigma_N)x + \bar{\mu}(0)$ with $j$ being the current density flowing in N, yielding a current carrying state.

The solution of Eq. (4.136) has the form

$$\delta\mu(x) = \alpha_1 e^{x/\lambda_S} + \alpha_2 e^{-x/\lambda_S}, \qquad (-d/2 < x < d/2) \tag{4.137}$$

where the coefficients $\alpha_1$ and $\alpha_2$ are determined by the spin-dependent tunnel currents $I_{i\sigma}$ passing through junctions 1 and 2

$$I_{1\sigma} = \frac{G_{1\sigma}}{e}\left[eV_1 - \sigma\delta\mu(-d/2)\right], \qquad I_{2\sigma} = \frac{G_{2\sigma}}{e}\left[eV_2 + \sigma\delta\mu(d/2)\right], \tag{4.138}$$

where $G_{i\sigma}$ is the tunnel conductance for electrons with spin $\sigma$ at junction $i$ ($i = 1, 2$) and is given by $G_{i\sigma} \propto \mathcal{D}_N \mathcal{D}_{Fi}^\sigma$, with $\mathcal{D}_N$ and $\mathcal{D}_{Fi}^\sigma$ being the densities of states of N and FM$i$ at the Fermi level, respectively, and $V_1$ and $V_2$ are the voltage drops at junction 1 and 2, respectively. The spin polarized tunneling is characterized by the following quantities:

$$P_1 = \frac{G_{1\uparrow} - G_{1\downarrow}}{G_{1\uparrow} + G_{1\downarrow}} = \frac{\mathcal{D}_{F1}^\uparrow - \mathcal{D}_{F1}^\downarrow}{\mathcal{D}_{F1}^\uparrow + \mathcal{D}_{F1}^\downarrow}, \quad \tilde{P}_2 = \frac{G_{2\uparrow} - G_{2\downarrow}}{G_{2\uparrow} + G_{2\downarrow}} = \frac{\mathcal{D}_{F2}^\uparrow - \mathcal{D}_{F2}^\downarrow}{\mathcal{D}_{F2}^\uparrow + \mathcal{D}_{F2}^\downarrow}, \tag{4.139}$$

where $\tilde{P}$ changes its sign depending on whether the magnetizations are parallel or antiparallel; $\tilde{P}_2$ is denoted by $P_2$ for the parallel alignment and by $-P_2$ for the antiparallel alignment. If the tunneling matrix elements is spin-independent, then $P_i$ is expressed in terms of the densities of states of as in the third terms.

In the following, we consider the case that the voltage drop in N is much smaller than those in the barrier: $(I_{inj}/A\sigma_N)d \ll V$ and $V \approx V_1 + V_2$, implying $R_iA/\rho_N d \gg 1$ ($i = 1, 2$), where $A$ is the junction area and $R_i = 1/(G_{i\uparrow} + G_{i\downarrow})$ the resistance of the $i$th junction. The resistance of the system is dominated by the barrier resistances, and the small voltage drops in the FM and N electrodes are neglected. It should be noted that so long as $R_iA/\rho_N d \gg 1$, the spin accumulation in the FMs near the junctions has little effect on the transport properties of FM1/N/FM2. However, when $R_iA/\rho_N d$ is comparable to or less than unity, one has to take into account the spin accumulation in the FMs as in the case of metallic multilayers [76].

When electrons tunnel through the barrier without spin-flip scattering, the tunnel current carried by spin up (down) electrons at the $i$th junction is equal to the corresponding current at the N side of the interface ($x = \pm d/2$), which leads to the boundary conditions for the spin current: $I_{1\uparrow} - I_{1\downarrow} = Aj_{spin}(x)|_{x=-d/2}$ and $I_{2\uparrow} - I_{2\downarrow} = Aj_{spin}(x)|_{x=d/2}$. The conservation of the charge current at the junctions gives another boundary condition: $I_{inj} = I_{1\uparrow} + I_{1\downarrow} = I_{2\uparrow} + I_{2\downarrow}$. These boundary conditions are used to determine $\delta\mu(x)$, $V_1$, and $V_2$.

### 4.4.1.1 Tunnel junctions with symmetric barriers

In the symmetric tunnel junctions, the tunnel barriers are identical and the electrodes are the same ferromagnet, so that the junction resistances, $R_1$ and

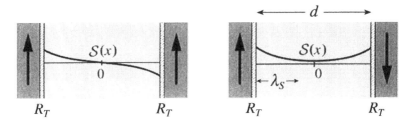

Figure 4.34: Schematic representation of the spin density $\mathcal{S}(x)$ in N created by the spin-polarized tunnel currents from the ferroamgnets for the parallel (left) and antiparallel (right) alignments of magnetizations in the FM electrodes.

$R_2$, as well as the spin polarizations, $P_1$ and $P_2$, have the same values; $R_1 = R_2 = R_T$ and $P_1 = P_2 = P$. This symmetric case is particularly useful to understand how the spin polarized electrons are injected and accumulated in N of the double barrier junctions and how the spin accumulation depends on the relative direction of magnetizations. It is obvious that the voltage drop at each junction is the same and equal to $V/2$, i.e., $V_1 = V_2 = V/2$. The shifts of the chemical potentials $\delta\mu_F(x)$ and $\delta\mu_A(x)$ for the parallel and antiparallel alignments are odd and even functions of $x$, respectively. The corresponding spin densities $\mathcal{S}_F(x)$ and $\mathcal{S}_A(x)$ are given by

$$\mathcal{S}_F(x) = \frac{-\frac{1}{2}\mathcal{D}_N PeV}{1 + \Lambda_S \coth{(d/2\lambda_S)}} \left[ \frac{\sinh{(x/\lambda_S)}}{\sinh{(d/2\lambda_S)}} \right], \qquad (4.140)$$

$$\mathcal{S}_A(x) = \frac{\frac{1}{2}\mathcal{D}_N PeV}{1 + \Lambda_S \tanh{(d/2\lambda_S)}} \left[ \frac{\cosh{(x/\lambda_S)}}{\cosh{(d/2\lambda_S)}} \right], \qquad (4.141)$$

with the non-dimensional parameter

$$\Lambda_S = \frac{R_T A}{\rho_N \lambda_S},$$

where $\rho_N = 1/\sigma_N$ is the resistivity of the normal metal. The quantity $R_T A$ is the specific resistance per unit area of the junction, having units of $\Omega\text{cm}^2$ or $\Omega\mu\text{m}^2$. The tunnel resistances of the double junction for the parallel and antiparallel alignments are given by

$$\begin{aligned} R_F &= 2R_T \frac{1 - P^2 + \Lambda_S \coth{(d/2\lambda_S)}}{1 + \Lambda_S \coth{(d/2\lambda_S)}}, \\ R_A &= 2R_T \frac{1 - P^2 + \Lambda_S \tanh{(d/2\lambda_S)}}{1 + \Lambda_S \tanh{(d/2\lambda_S)}}, \end{aligned} \qquad (4.142)$$

which yields the TMR ratio defined by $\Delta R/R_F$ with $\Delta R = (R_A - R_F)$:

$$\frac{\Delta R}{R_F} = \frac{P^2 \cosh^{-2}{(d/2\lambda_S)}}{\left[1 - P^2 + \Lambda_S \tanh{(d/2\lambda_S)}\right]\left[1 + \Lambda_S^{-1} \tanh{(d/2\lambda_S)}\right]}. \qquad (4.143)$$

Figure 4.35 shows the spin-injection signal $\Delta R$ and TMR as functions of reduced thickness $d/\lambda_S$ for different values of $\Lambda_S = (R_T A)/(\rho_N \lambda_S)$. The curves of $\Delta R$ are almost the same to those obtained by Fert and Lee [78], Hershfield and Zhao [77], who calculated $\Delta R$ based on the Johnson's spin injection device.

**Thick film** $(d \gg 2\lambda_S)$  When the thickness of the normal metal is much larger than the spin-diffusion length $(d \gg 2\lambda_S)$, the spin density decays with $\lambda_S$ from the junctions and vanishes in the interior of N. From Eqs. (4.140) and (4.141), the spin density created near junction 1 is expressed as

$$\mathcal{S}_F(x) = \mathcal{S}_A(x) \approx \frac{\mathcal{D}_N PeV}{2\Lambda_S} \exp\left[-\left(x + d/2\right)/\lambda_S\right], \quad (-d/2 < x < 0) \quad (4.144)$$

for both alignments. In this case, the spin densities penetrated from the junctions are nearly decoupled from each other, and the TMR shows an exponential decay: TMR $\sim \exp(-d/\lambda_S)$.

A measure of the spin injection efficiency in the thick film is given by the value of $\mathcal{S}$ at the junction interfaces $(x = \pm d/2)$:

$$\mathcal{S}_F(d/2) = \mathcal{S}_A(d/2) = \frac{\frac{1}{2}\mathcal{D}_N PeV}{1 + \Lambda_S} \approx \frac{P\lambda_S}{2eD} \frac{I_{\text{inj}}}{A}. \quad (4.145)$$

The magnitude of the accumulated spin density depends on $\Lambda_S$. For good metals like Al and Au, the resistivity is $\rho_N < 10^{-6}\,\Omega$cm and the spin diffusion length $\lambda_S \sim 10^{-3}$ cm. The value of the specific resistance $R_T A$ of tunnel junctions that exhibit a large MR is widely spread, ranging from $\sim 10^{-7}\,\Omega$cm$^2$ to $\sim 10^{-2}\,\Omega$cm$^2$. If these values are used to estimate the value of $\Lambda_S$, one finds $\Lambda_S > 10^2$, so that the injected spin density is very small compared with the optimum value of $\mathcal{D}_N PeV/2$ for $\lambda_S \to \infty$. Therefore, the spin injection efficiency becomes very small unless one uses the junctions of low specific resistance.

**Thin film** $(d \ll 2\lambda_S)$  When the thickness of the normal layer is much smaller than the spin-diffusion length, the efficiency of spin injection is very much enhanced. For $d \ll 2\lambda_S$, Eq. (4.135) with Eqs. (4.140) and (4.141) is reduced to

$$\mathcal{S}_F(x) = \frac{\frac{1}{2}\mathcal{D}_N PeV}{1 + (2\lambda_S/d)\,\Lambda_S}\left(-\frac{2x}{d}\right) \approx 0, \quad (4.146)$$

$$\mathcal{S}_A(x) = \frac{\frac{1}{2}\mathcal{D}_N PeV}{1 + (d/2\lambda_S)\,\Lambda_S}. \quad (4.147)$$

We see that the additional factors $(2\lambda_S/d)$ and $(d/2\lambda_S)$ newly appear in the denominators in Eqs. (4.146) and (4.147) in the thin film case $(d/\lambda_S \ll 1)$.

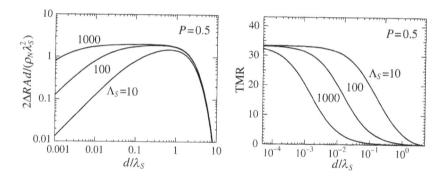

Figure 4.35: Spin-injection signal $\Delta R$ (left) and TMR (right) vs $d/\lambda_S$ for different values of $\Lambda_S = R_T A/(\rho_N \lambda_S)$ in the case of symmetric junction.

These factors, which are absent in the thick case (see Eq. (4.144)), originates from the interference effect between the penetrated spin densities from the two sides; the spin densities interfere destructively in the parallel alignment, while constructively in the antiparallel alignment.

The TMR in the thin film limit $(d/2\lambda_S \ll 1)$ is given by

$$\frac{\Delta R}{R_F} \approx \frac{P^2}{1 - P^2 + \Gamma_S}, \tag{4.148}$$

where $\Gamma_S = (d/2\lambda_S)\Lambda_S = (R_T Ad/2\rho_N \lambda_S^2)$. If $\Gamma_S < 1$ is fulfilled, the spin density is given by $\mathcal{S}_F \simeq 0$ and $\mathcal{S}_A \approx \frac{1}{2}\mathcal{D}_N PeV$, indicating that a nearly full value of spin density is accumulated in N for the antiparallel alignment while a negligibly small spin density for the parallel alignment, yielding the optimum TMR ratio $TMR = P^2/(1 - P^2)$. If the typical values of $\rho_N \sim 10^{-6}\,\Omega\mathrm{cm}$, $d \sim 10^{-6}\,\mathrm{cm}$ and $\lambda_S \sim 10^{-3}\,\mathrm{cm}$ are used, the condition $\Gamma_S < 1$ is expressed as $R_T A < 10^{-6}\,\Omega\mathrm{cm}^2$, which corresponds to $R_T < 100\,\Omega$ for the junction with area $1\,\mu\mathrm{m}^2$.

The spin current flowing through N is given by

$$j_{\mathrm{spin}}^F = P\frac{V}{2R_T A}, \qquad j_{\mathrm{spin}}^A \approx 0. \tag{4.149}$$

It is interesting to note that, in the F alignment the spin density is not accumulated in N but the spin current flows through N, while in the A alignment the spin density is accumulated but the spin current is absent in the thin film case $(d/2\lambda_S \ll 1)$.

### 4.4.1.2   Tunnel junctions with asymmetric barriers

In the asymmetric tunnel junctions, both the tunnel barriers and the ferromagnets of the electrodes are different; $R_1 \neq R_2$ and $P_1 \neq P_2$. In the following, we restrict ourselves to the case where the thickness of N is much smaller than the spin-diffusion length, i.e., $d \ll 2\lambda_S$. In this case, we have the spin densities

$$\mathcal{S}_F = \left[ \frac{r_1 r_2 (P_1 - P_2)}{1 - (r_2 P_1 + r_1 P_2)^2 + \Gamma_S} \right] \mathcal{D}_N e V, \qquad (4.150)$$

$$\mathcal{S}_A = \left[ \frac{r_1 r_2 (P_1 + P_2)}{1 - (r_2 P_1 - r_1 P_2)^2 + \Gamma_S} \right] \mathcal{D}_N e V, \qquad (4.151)$$

with $r_i$ the reduced junction resistance of $i$th junction

$$r_1 = R_1 / (R_1 + R_2), \qquad r_2 = R_2 / (R_1 + R_2), \qquad (r_1 + r_2 = 1)$$

and $\Gamma_S$ the spin relaxation parameter

$$\Gamma_S = r_1 r_2 (\tau_t / \tau_S) = \frac{(d/\lambda_S)^2}{(\rho_N d / R_1 A) + (\rho_N d / R_2 A)}, \qquad (4.152)$$

where $\tau_t = 2e^2 \mathcal{D}_N (R_1 + R_2) A d$ is the characteristic tunneling time. The reduced tunnel resistance is connected with the reduced tunnel conductance $g_i = G_i / (G_1 + G_2)$ as $r_1 = g_2$ and $r_2 = g_1$. We note that the spin densities in N for the parallel and antiparallel alignments are proportional to the difference $P_1 - P_2$ and the sum $P_1 + P_2$ of the spin polarizations at the electrodes, respectively, irrespective of the asymmetry of the tunnel barriers. The voltage drops at junctions 1 and 2 are given by

$$V_1 = r_1 V + \delta V, \qquad V_2 = r_2 V - \delta V, \qquad (4.153)$$

where the first terms represent the Ohmic voltage and the second terms the deviation $\pm \delta V$ due to the spin accumulation:

$$\delta V = \frac{r_1 r_2 (r_2 P_1 + r_1 \tilde{P}_2)}{1 - (r_2 P_1 + r_1 \tilde{P}_2)^2 + \Gamma_S} (P_1 - \tilde{P}_2) V. \qquad (4.154)$$

Note that the strong mixing between up and down spins ($\Gamma_S \gg 1$) gives a simple Ohmic law $V_i = r_i V$.

The resistance $R = V/I$ of the double junction is calculated as

$$R = (R_1 + R_2) \left[ \frac{1 - (r_2 P_1 + r_1 \tilde{P}_2)^2 + \Gamma_S}{1 - (r_2 P_1^2 + r_1 P_2^2) + \Gamma_S} \right]. \qquad (4.155)$$

The spin-injection signal $\Delta R = R_A - R_F$ is given by

$$\Delta R = (R_1 + R_2) \frac{4 r_1 r_2 P_1 P_2}{1 - (r_2 P_1^2 + r_1 P_2^2) + \Gamma_S}. \qquad (4.156)$$

From Eqs. (4.155) and (4.156), the TMR is obtained as

$$\frac{\Delta R}{R_F} = \frac{4r_1 r_2 P_1 P_2}{1 - (r_2 P_1 + r_1 P_2)^2 + \Gamma_S}. \tag{4.157}$$

The TMR is degraded in the case of strong asymmetry in the conductances ($R_1 \ll R_2$ or $R_1 \gg R_2$). A large TMR ratio is obtained when the following conditions are satisfied; the tunnel barriers are similar ($R_1 \sim R_2$) and the spin relaxation time in N is long compared with the tunneling time ($\Gamma_S < 1$). The latter condition is $(\rho_N/R_1) + (\rho_N/R_2) > (Ad/\lambda_S^2)$, which requires a low junction resistance and/or a thin N with $d$ much smaller than $\lambda_S$. If these conditions are satisfied, we have the optimum ratio $\Delta R/R_F \sim P_1 P_2/(1 - P_1 P_2)$ in the normal state.

It is interesting to note that the resistance difference $\Delta R$ for $\Gamma_S \ll 1$ is scaled with the junction parameters as

$$\Delta R = \frac{4r_1 r_2 P_1 P_2}{1 - (r_2 P_1^2 + r_1 P_2^2)}(R_1 + R_2), \tag{4.158}$$

while $\Delta R$ for $\Gamma_S \gg 1$ is scaled with the parameters of the normal metal

$$\Delta R = 4P_1 P_2 \left( \frac{\rho_N \lambda_S^2}{Ad} \right). \tag{4.159}$$

The latter result reminds us of the classic result $\Delta R = -4P_1 P_2 (\rho_N \lambda_S^2/Ad)$ of the spin-coupled impedance in the three terminal device [79] shown in Fig. 4.33, in which the current enters through FM1 and exits through N, and FM2 is a voltage probe for measuring the non-equilibrium spin density in N. Note that the sign of $\Delta R$ in Eq. (4.159) is opposite from that of the Johnson's device.

## 4.4.2    Spin-dependent Coulomb blockade in a FM/N/FM junction

In Sec. 4.3, we studied the effect of CB on charge transport in various types of magnetic nanostructures. In this section, we discuss the spin-dependent CB in an SET consisting of a very small normal-metal island between two ferromagnetic electrodes. The circuit diagram is the same as that in Fig. 4.20 (a), except that the central island is a normal metal (N). In the F/N/F double tunnel junction, the CB is influenced by spin accumulation in N, depending on the relative orientation of magnetizations in the left and right hand electrodes. In the antiferromagnetic (A) alignment, it is easy (difficult) for the up-spin (down-spin) electrons to tunnel into the normal metal, while it is difficult (easy) to tunnel out of it, because the density of states for each spin band is different between the left and right electrodes. This imbalance among the tunnel currents causes the non-equilibrium spin accumulation [74, 76, 79], when

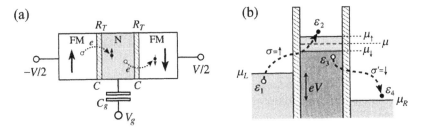

Figure 4.36: (a) Single electron transistor (SET) consisting of a normal-metal island sandwiched between two ferromagnets and a capacitively coupled gate. (b) Energy diagram illustrating a tunneling process in SET involved in spin-dependent cotunneling: An electron with up-spin ($\sigma =\uparrow$) tunnels to the island N through the 1st junction, and the virtual intermediate state decays by a different electron with down-spin ($\sigma' =\downarrow$) tunneling through the 2nd junction, and vice versa.

the spin relaxation time is sufficiently long in N. In the CB region (Fig. 4.21), the spin accumulation is caused by cotunneling, since the sequential tunneling is suppressed by CB at low temperatures ($k_B T \ll E_c$). The spin accumulation increases (decreases) the chemical potential of the up-spin (down-spin) electrons by $\delta\mu_\uparrow = \mu_\uparrow - \mu$ ($\delta\mu_\downarrow = \mu_\downarrow - \mu$), where $\mu_\uparrow$ and $\mu_\downarrow$ are the chemical potentials of the up- and down-spin electrons, respectively. This shift of the chemical potential $\delta\mu_\sigma$ decreases the energy for adding (extracting) an down-spin (up-spin) electron to (from) the N island. In the A alignment, the spin-splitting by spin accumulation plays the role of spin-dependent gate voltage, and therefore removes the spin-degeneracy of the boundary of the CB region as shown above.

We consider, for simplicity, a symmetric SET with the same ferromagnets and the identical insulating barriers as shown Fig. 4.36(a). We calculate the cotunneling current by taking into account the spin-splitting of the chemical potentials in the island, in which the spin accumulation is described by the Fermi function of the form $f(\epsilon - \delta\mu_\sigma)$.

The cotunneling current through the double junction is expressed as

$$I = e \sum_{n=-\infty}^{\infty} \sum_{\sigma,\sigma'=\uparrow\downarrow} p(n) \left[ I_{1\sigma 2\sigma'}(n) - I_{2\sigma'1\sigma}(n) \right], \qquad (4.160)$$

where $n$ is the number of excess electrons in the normal-metal island, $p(n)$ is the probability of charge state $n$, and $I_{i\sigma j\sigma'}(n)$ represents the current due to the tunneling process in which an electron with spin-$\sigma$ tunnels into the central island through the $i$th junction and another electron with spin $\sigma'$ tunnels out of it through the $j$th junction. For example, $I_{1\uparrow 2\downarrow}(n)$ is contributed from the co-tunneling processes shown in Fig. 4.36(b). The probability $p(n)$ is determined by the condition for detailed balancing:

$$p(n) \sum_\sigma \left[ \Gamma_{1\sigma}^+(n) + \Gamma_{2\sigma}^-(n) \right] = p(n+1) \sum_\sigma \left[ \Gamma_{1\sigma}^-(n+1) + \Gamma_{2\sigma}^-(n+1) \right],$$

where $\Gamma_{i\sigma}^\pm(n)$ is the tunneling rate for $n \to n \pm 1$ across the $i$th junction, and is calculated from the golden rule (see Eq. (4.74)) by taking into account the spin-splitting of the chemical potentials in N:

$$\Gamma_i^\pm(n) = \left( \frac{1}{e^2 R_{i\sigma}} \right) \frac{E_{i\sigma}^\pm(n)}{\exp\left[ E_{i\sigma}^\pm(n)/k_B T \right] - 1}. \tag{4.161}$$

Here, $R_{i\sigma}$ is the tunnel resistance for electrons with spin $\sigma$ across the $i$-th junction and $E_{i\sigma}^\pm(n)$ is the change in the electrostatic energy associated with tunneling of each process

$$E_{i\sigma}^+(n) = E_i^+(n) + \delta\mu_\sigma, \qquad E_{i\sigma}^-(n) = E_i^-(n) - \delta\mu_\sigma, \tag{4.162}$$

where $E_i^\pm(n)$ is given in Eqs. (4.94) and (4.96).

The current $I_{i\sigma j\sigma'}(n)$ is calculated in the same way as in Sec. 4.3.3, thereby replacing the Fermi function $f(\epsilon)$ in N with $f(\epsilon - \delta\mu_\sigma)$ in Eq. (4.100). The result is

$$I_{i\sigma j\sigma'}(n) = \frac{R_K}{4\pi^2 e R_{i\sigma} R_{j\sigma'}} \int_{-\infty}^\infty d\epsilon \, \frac{\epsilon(\epsilon - V_{\sigma\sigma'}^{ij}) \exp[\beta e V_{\sigma\sigma'}^{ij}/2]}{\sinh[\frac{1}{2}\beta\epsilon] \sinh[\frac{1}{2}\beta(\epsilon - V_{\sigma\sigma'}^{ij})]} |M_{\sigma\sigma'}|^2, \tag{4.163}$$

where $R_K = h/e^2$ is the resistance quantum, $V_{\sigma\sigma'}^{ij} = (j-i)V - \delta\mu_\sigma + \delta\mu_{\sigma'}$ and $M_{\sigma\sigma'}$ is given by

$$M_{\sigma\sigma'} = \frac{1}{\epsilon + E_{i\sigma}^+(n) + i\gamma_+} + \frac{1}{-\epsilon + eV_{\sigma\sigma'} + E_{j\sigma'}^-(n) + i\gamma_-}, \tag{4.164}$$

with $\gamma_\pm = (\hbar/2e^2) \sum_\sigma \sum_i [E_{i\sigma}^\pm(n)/R_{i\sigma}] \coth[E_{i\sigma}^\pm(n)/2k_B T]$. The tunnel resistances for the majority and minority spin bands are taken to be $R_M = 2R_K$ and $R_m = [(1+P)/(1-P)]R_M$, respectively, and $1/R_T = 1/R_M + 1/R_m$.

The chemical potential shift $\delta\mu \equiv \delta\mu_\uparrow = -\delta\mu_\downarrow$ is determined by the stationary condition that the amount of incoming spins are balanced with that of outgoing spins in N

$$\sum_{n=-\infty}^\infty \sum_{i,j=1}^2 p(n) \left[ I_{i\uparrow j\downarrow}(n) - I_{i\downarrow j\uparrow}(n) \right] = 0. \tag{4.165}$$

In the F alignment, there is no spin accumulation in N, because $\delta\mu = 0$ ensures the detail balancing $I_{i\uparrow j\downarrow}(n) = I_{i\downarrow j\uparrow}(n)$. Therefore the CB region is the same as that for usual metallic single electron transistors as shown in Fig. 4.21. In

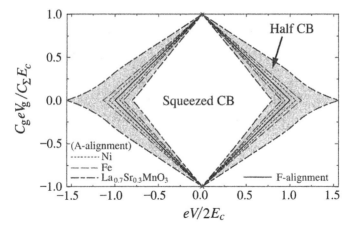

Figure 4.37: Boundary of the CB region for the ferromagnetic (F) alignment is indicated by the solid line. The boundaries of the half CB region for the antiferromagnetic (A) alignment with Ni, Fe, and $La_{0.7}Sr_{0.3}MnO_3$ electrodes are plotted by dotted, dashed, and dot-dashed lines, respectively. The half CB region for $La_{0.7}Sr_{0.3}MnO_3$ electrodes is indicated by the shading. The tunnel resistance for the majority spin band is taken to be $R_M = 2R_K$ for all systems.

the A alignment, however, the spins are accumulated, because the stationary condition is violated for $\delta\mu = 0$ at a finite bias $V$, i.e., $I_{i\uparrow j\downarrow}(n) \neq I_{i\downarrow j\uparrow}(n)$, and gives rise to a non-zero shift of $\delta\mu$. To obtain $\delta\mu$, we carry out the numerical integration of Eq. (4.163).

There are eight different processes of a single electron tunneling in the SET of FM/N/FM, each of which is accompanied by the change in the electrostatic energy $E_{i\sigma}^{\pm}$. The CB region in the A alignment is determined by the condition that all single electron tunneling processes increases the charging energy of the island, i.e., $E_{1\sigma}^{\pm} > 0$ and $E_{2\sigma}^{\pm} > 0$. In the A alignment, the spin accumulation removes the spin-degeneracy in the CB region. The inner boundary, which is determined by $E_{1\downarrow}^{+} = 0$, $E_{1\uparrow}^{-} = 0$, $E_{2\uparrow}^{+} = 0$, and $E_{2\downarrow}^{-} = 0$, defines the squeezed CB region as shown in Fig. 4.37. Outside the squeezed CB region, we have a new anomalous region indicated by shading, in which the up-spin (down-spin) electrons can tunnel into (out of) the normal-metal island, whereas the spin-down (spin-up) electrons cannot. This is "half CB" region which is given by the condition $-\delta\mu < E_1^{\pm}(n) < \delta\mu$ and $-\delta\mu < E_2^{\pm}(n) < \delta\mu$. In Fig. 4.37, the boundaries of this half CB region for $n = 0$ in the A alignment of Ni, Fe and $La_{0.7}Sr_{0.3}MnO_3$ electrodes are plotted. The solid line indicates the CB boundary for the F alignment and the shaded area represents the half CB region for the A alignment with $La_{0.7}Sr_{0.3}MnO_3$ electrodes in Fig. 4.37. One can see that the half CB region increases and the squeezed CB region, which

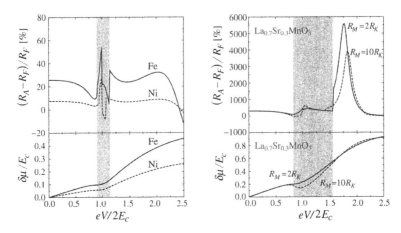

Figure 4.38: Left: Tunnel magnetoresistance for Ni and Fe electrodes are plotted against the bias voltage $V$ in the upper panel. The chemical potential shift $\delta\mu$ is plotted in the lower panel. The shaded area represents the half CB region for Fe electrodes. The tunnel resistance for the majority spin electrons is $R_M = 2R_K$, the gate voltage is $V_g = 0$ and $T = 0$. Right: The same plot for $La_{0.7}Sr_{0.3}MnO_3$ electrodes. The solid(dashed) lines indicate the results for $R_M = 2R_K(10R_K)$. The half Coulomb blockade region for $R_M = 2R_K$ is indicated by the shading. The gate voltage is $V_g = 0$ and $T = 0$.

is surrounded by the half CB region, decreases as $P$ increases, because $\delta\mu$ increases with $P$.

Figure 4.38 shows the TMR, $(R_A - R_F)/R_F$, and the shift of the chemical potential $\delta\mu$ at $T = 0$ and $V_g = 0$. At $V \simeq 0$, the TMR for Ni, Fe and $La_{0.7}Sr_{0.3}MnO_3$ electrodes are about 7.5%, 26% and 310%, respectively. The values are $35 \sim 40\%$ larger than those in the absence of CB. As the bias voltage $V$ increases from 0 toward the boundary of the squeezed CB region, the tunnel current of the A alignment increases more rapidly than that of the F alignment and the TMR decreases. At the boundaries of the half CB region, $R_A$ jumps because the current due to cotunneling decreases rapidly and that due to sequential tunneling starts to flow. The same jump in $R_F$ appears at $eV/2E_c = 1$, which is the boundary of the CB region for the F alignment. Therefore, the TMR oscillates in the half CB region as shown in Fig. 4.38. We also find that cotunneling enhances the TMR around $eV/2E_c \simeq 2.0$, because it suppresses the $V$ dependence of the total current for the A alignment. In order to see what happens when cotunneling is suppressed, we calculate the TMR in the system with large tunnel resistance, $R_M = 10R_K$, In Fig. 4.38, TMR and $\delta\mu$ for the electrodes of $La_{0.7}Sr_{0.3}MnO_3$ and junctions with $R_M = 10R_K$ are plotted by the dashed lines. The size of the half CB region is 79% of that for $R_M = 2R_K$.

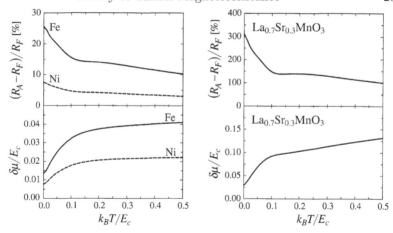

Figure 4.39: Left: The temperature dependence of the tunnel magnetoresistance at $V_g = 0$ and $eV/2E_c = 0.1$ for Ni and Fe electrodes is shown in the upper panel. The temperature dependence of the chemical potential shift $\delta\mu$ at $V_g = 0$ and $eV/2E_c = 0.1$ for Ni and Fe electrodes is plotted in the lower panel. Right: The same plot for $La_{0.7}Sr_{0.3}MnO_3$ electrodes. In both figures the tunnel resistance for the majority spin electrons is $R_M = 2R_K$.

The temperature dependencies of TMR and $\delta\mu$ have been calculated at $V_g = 0$ and $eV/2E_c = 0.1$ in the squeezed CB region. As temperature $T$ increases, the sequential tunneling, which is exponentially suppressed at low $T$, is recovered. The results are shown in Fig. 4.39. One can see clearly the crossover between cotunneling and sequential tunneling near $k_B T/E_c = 0.1$ [65].

## 4.4.3 Spin injection into a superconductor

The spin-polarized current driven from a ferromagnet (FM) into a superconductor (SC) creates a non-equilibrium spin polarization in N [74] or SC [73, 80]. Recently, there has been a number of experiments demonstrating that superconductivity is strongly suppressed by injection of spin-polarized electrons using tunnel junctions consisting of a high-$T_c$ SC and a ferromagnetic manganite [81-85].

In this section, we discuss non-equilibrium spin imbalance created by a spin-polarized tunnel current and their competition with superconducting condensate in a SC sandwiched between two ferromagnets (FM1/SC/FM2) [86, 87]. A particular emphasis is placed upon spin-dependent effect, i.e., dependence of the transport properties on the relative orientation of the magnetizations in the FM electrodes. In the following, taking account of asymmetry in the junction, we describe how the spin density is accumulated in SC and suppresses

the superconductivity of SC, depending on the relative orientation of magnetizations as well as the difference in the tunnel resistance of the barriers and
the spin polarization of FMs.

We consider a FM1/SC/FM2 double tunnel junction, which is the same as
that in Fig. 4.33(a), if one replaces N with SC. The left and right electrodes
are made of different ferromagnets and the central one is a superconductor
with thickness $d$. The magnetization of FM1 is chosen to point up and that
of FM2 is either up or down. In the asymmetric tunnel junction, the height
of the tunnel barriers and/or the strength of the ferromagnets are different,
which are characterized by the different values of the tunnel resistance and
those of the spin-polarization in the junction.

We calculate the tunnel current using a phenomenological tunnel Hamiltonian. If SC is in the superconducting state, it is convenient to rewrite the
electron operators $a_{k\sigma}$ in SC in terms of the quasiparticle operators $\gamma_{k\sigma}$ using
the Bogoliubov transformation [88]

$$a_{k\uparrow} = u_k \gamma_{k\uparrow} + v_k^* \gamma_{-k\downarrow}^\dagger, \qquad a_{-k\downarrow}^\dagger = -v_k \gamma_{k\uparrow} + u_k^* \gamma_{-k\downarrow}^\dagger,$$

where $|u_k|^2 = 1 - |v_k|^2 = \frac{1}{2}(1 + \xi_k/E_k)$ with the quasiparticle dispersion
$E_k = \sqrt{\xi_k^2 + \Delta^2}$ of SC, $\xi_k$ being the one-electron energy relative to the chemical
potential and $\Delta$ being the gap parameter. Then, using the golden rule formula,
we obtain the spin-dependent currents $I_{i\sigma}$ across the $i$th junction [65]:

$$I_{1\uparrow} = (G_{1\uparrow}/e\mathcal{D}_N)\,[\mathcal{N}_1 - \mathcal{S}]\,, \quad I_{1\downarrow} = (G_{1\downarrow}/e\mathcal{D}_N)\,[\mathcal{N}_1 + \mathcal{S}]\,, \quad (4.166)$$
$$I_{2\uparrow} = (G_{2\uparrow}/e\mathcal{D}_N)\,[\mathcal{N}_2 + \mathcal{S}]\,, \quad I_{2\downarrow} = (G_{2\downarrow}/e\mathcal{D}_N)\,[\mathcal{N}_2 - \mathcal{S}]\,, \quad (4.167)$$

where $G_{i\sigma}$ $(i = 1, 2)$ is the tunnel conductance of the $i$th junction for electrons
with spin $\sigma$ if SC is in the normal state, and is given by $G_{i\sigma} \propto |T_i^\sigma|^2 \mathcal{D}_N \mathcal{D}_{Fi}^\sigma$,
where $|T_i^\sigma|^2$ is the tunneling probability of the $i$th junction and $\mathcal{D}_N$ and $\mathcal{D}_{Fi}^\sigma$
are the spin-subband densities of states in SC and FM$i$, respectively. Here,
the nonequiribrium charge imbalance [89, 90] is neglected, because it has little
relevance to the spin-dependent effect [91]. The quantity $\mathcal{N}_i$ is given by [92]

$$\mathcal{N}_i = \frac{1}{2}\sum_k \Big[ f_0\big(E_k - eV_i\big) - f_0\big(E_k + eV_i\big) \Big], \qquad (4.168)$$

where $f_0$ is the Fermi distribution function of thermal equilibrium in FM and
$V_i$ the voltage drop at the $i$th junction ($V_1 + V_2 = V$). The quantity $\mathcal{S}$ is
quasiparticle spin density in SC and are defined by

$$\mathcal{S} = \frac{1}{2}\sum_k \big(f_{k\uparrow} - f_{k\downarrow}\big)\,, \qquad (4.169)$$

where $f_{k\sigma} = \langle \gamma_{k\sigma}^\dagger \gamma_{k\sigma} \rangle$ is the distribution function of quasiparticles with energy
$E_k$ and spin $\sigma$ in the state $k$.

The conservation of total charge $Q_{\text{tot}} = \sum_{\mathbf{k}\sigma} \langle a_{\mathbf{k}\sigma}^\dagger a_{\mathbf{k}\sigma} \rangle$ in SC gives $I_{1\uparrow} + I_{1\downarrow} = I_{2\uparrow} + I_{2\downarrow}$, which yields the relation

$$\left( g_1 P_1 + g_2 \tilde{P}_2 \right) \mathcal{S} = g_1 \mathcal{N}_1 - g_2 \mathcal{N}_2, \tag{4.170}$$

where $g_i = G_i/(G_1 + G_2)$ $(g_1 + g_2 = 1)$ is the reduced conductance of the $i$th junction, and

$$P_1 = \frac{G_{1\uparrow} - G_{1\downarrow}}{G_{1\uparrow} + G_{1\downarrow}}, \qquad \tilde{P}_2 = \frac{G_{2\uparrow} - G_{2\downarrow}}{G_{2\uparrow} + G_{2\downarrow}},$$

where $\tilde{P}_2 = P_2$ for the F alignment and $\tilde{P}_2 = -P_2$ for the A alignment of magnetizations. $P_1$ and $P_2$ are the degree of spin-polarization of FM1 and FM2.

The quasiparticle spin density $\mathcal{S}$ created in SC is calculated by balancing the spin injection rate $(d\mathcal{S}/dt)_{\text{inj}} = [(I_{1\uparrow} - I_{1\downarrow}) - (I_{2\uparrow} - I_{2\downarrow})]/2e$ with the spin relaxation rate $\mathcal{S}/\tau_S$, where $\tau_S$ is the spin-relaxation time. The result is combined with Eq. (4.170) to yield

$$\mathcal{S} = \frac{g_1 g_2 (P_1 - \tilde{P}_2)}{1 - (g_1 P_1 + g_2 \tilde{P}_2)^2 + \Gamma_S} (\mathcal{N}_1 + \mathcal{N}_2), \tag{4.171}$$

where $\Gamma_S = g_1 g_2 (\tau_t/\tau_S)$, $\tau_t = 2e^2 \mathcal{D}_N (R_1 + R_2) A d$ is the tunneling time, and $A$ the junction area, and $R_j = 1/G_j$. Note that the spin density in SC is proportional to the *difference* $(P_1 - P_2)$ for the F alignment and the *sum* $(P_1 + P_2)$ for the A alignment.

The superconducting gap $\Delta$ in SC is determined by $f_{\mathbf{k}\sigma}$ through the BCS gap equation

$$\frac{1}{V_{\text{BCS}}} = \sum_{\mathbf{k}} \frac{1 - f_{\mathbf{k}\uparrow} - f_{\mathbf{k}\downarrow}}{E_{\mathbf{k}}}. \tag{4.172}$$

It follows from Eq. (4.171) that, if the junction is symmetric, $\mathcal{S}$ vanishes for the F alignment, while $\mathcal{S} \neq 0$ for the A alignment. In the asymmetric case, $\mathcal{S}$ becomes finite for both alignments. In the following, we restrict ourselves to the case $\tau_t \ll \tau_S$. In addition, the thickness $d$ of SC is much smaller than the spin diffusion length $\lambda_S = \sqrt{D\tau_S}$, $D$ being the diffusion constant, so that the distribution of quasiparticles is spatially uniform in SC. Then, the distribution function $f_{\mathbf{k}\sigma}$ is described by $f_0$, but the chemical potentials of the spin-up and spin-down quasiparticles are shifted oppositely by $\delta\mu$ from the equilibrium one to generate the spin density;

$$f_{\mathbf{k}\uparrow} = f_0(E_{\mathbf{k}} - \delta\mu), \qquad f_{\mathbf{k}\downarrow} = f_0(E_{\mathbf{k}} + \delta\mu). \tag{4.173}$$

We solve self-consistently Eqs. (4.171) and (4.172) with respect to $\Delta$, $\delta\mu$, and $V_i$, and obtain $\Delta$ and $\mathcal{S}$ as functions of $V$. The results are used to calculate

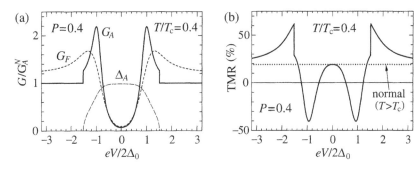

Figure 4.40: (a) Tunnel conductance as a function of bias voltage. The dashed and solid curves indicate the conductances $G_F$ and $G_A$ for the ferromagnetic and antiferromagnetic alignments, respectively, and the long-dashed curve the superconducting gap parameter $\Delta$ in the A alignment. (b) Tunnel magnetoresistance (TMR) as a function of bias voltage. The dotted line indicates TMR $= P^2/(1 - P^2)$ in the normal state.

the total tunnel current $I_{\text{inj}} = I_{i\uparrow} + I_{i\downarrow}$:

$$I_{\text{inj}} = \frac{1}{e\mathcal{D}_N} \left( \frac{\mathcal{N}_1 + \mathcal{N}_2}{R_1 + R_2} \right) \frac{1 - (g_1 P_1^2 + g_2 P_2^2) + \Gamma_S}{1 - (g_1 P_1 + g_2 \tilde{P}_2)^2 + \Gamma_S}, \tag{4.174}$$

which we call the injection current.

In the case of $P = P_1 = P_2$, the spin current $I_{\text{spin}} = I_{i\uparrow} - I_{i\downarrow}$ for the F alignment is given by

$$I_{\text{spin}} = P I_{\text{inj}}, \tag{4.175}$$

while for the A alignment

$$I_{\text{spin}} = \frac{(g_2 - g_1)(1 - P^2) \pm P\Gamma_S}{1 - (g_1 - g_2)^2 P^2 + \Gamma_S} I_{\text{inj}}, \tag{4.176}$$

where $I_{\text{spin}}$ with sign $\pm$ are the spin currents across junctions 1 and 2, respectively; the difference of which is equal to the rate of spin diffusion, $2e\mathcal{S}/\tau_S$, in SC as expected.

As the injection current $I_{\text{inj}}$ increases, the superconductivity is strongly suppressed by the pair breaking effect due to the increase of the quasiparticle spin density $\mathcal{S}$ in SC. The amount of $\mathcal{S}$ accumulated in SC is directly connected to the injection current $I_{\text{inj}}$ by the relation

$$\mathcal{S} = \left[ \frac{P_1 - \tilde{P}_2}{1 - (g_1 P_1^2 + g_2 P_2^2) + \Gamma_S} \right] \frac{e\mathcal{D}_N}{G_1 + G_2} I_{\text{inj}}. \tag{4.177}$$

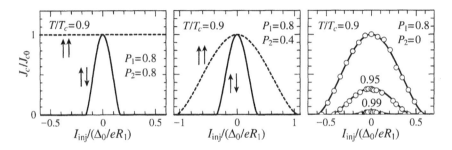

Figure 4.41: Dependence of the critical current $J_c$ on the injection current $I_{inj}$ for the spin polarization $P_1 = 0.8$ of FM1 and different values of $P_2$ of FM2. The open circles indicate the critical current measured at $T = 80$ K and $T = 84$ K ($T_c \sim 89$ K) by Dong *et al.* [82] in a FM/SC/N junction made of a high-$T_c$ SC and a ferromagnetic manganite with $P \sim 100\%$ [7].

When FM1 and FM2 are the same ($P_1 = P_2$), the injected spins vanish in the F alignment, and are accumulated only in the A alignment. Therefore the pair breaking effect occurs only in the A alignment. However, when the FMs are different, the injected spins are accumulated in proportion to ($P_1 - P_2$) for the F alignment and ($P_1 + P_2$) for the A alignment, and thus we have the pair breaking effect in both alignments.

Figure 4.40(a) shows the voltage dependence of the differential conductance $G_F$ and $G_A$ for the F and A alignments at $T/T_c = 0.4$ in the symmetric case; $P = P_1 = P_2$ and $R_T = R_1 = R_2$. The $G_F$ shows the ordinary dependence on $V$ expected for the constant gap $\Delta$. In contrast, because of the decrease in $\Delta_A$ ($\Delta$ in the A alignment is denoted by $\Delta_A$ hereafter) with increasing voltage, $G_A$ increases with voltage more rapidly than $G_F$, forming a higher peak than $G_F$, and then decreases steeply. At the critical voltage $V_c$, $G_A$ jumps to the conductance $G_A^N$ in the normal state. The TMR is calculated by the formula: TMR = $(G_F/G_A) - 1$. Using the values of Fig. 4.40(a), we obtain the $V$ dependence of TMR shown in Fig. 4.40(b). At $V = 0$, TMR takes the same value as in the normal state. A deep negative dip appears at $eV/2\Delta_0 \sim 1$ where $\Delta_A$ steeply decreases, exhibiting an inverse TMR effect ($G_A > G_F$), and is followed by the discontinuous jump at $V_c$ above which TMR is highly enhanced compared to that in the normal state.

The suppression of the superconducting gap $\Delta$ by spin injection is detected by measuring the superconducting critical current $J_c$ . According to the Ginzburg-Landau theory, $J_c$ is proportional to $\Delta^3$, because $J_c \propto \Delta^2 v_c$, $v_c$ being the critical superfluid velocity, and $v_c \propto \Delta$ [88]. Figure 4.41 shows the cube of the normalized gap, $(\Delta/\Delta_0)^3$, and thus the normalized critical current, $(J_c/J_{c0})$, as a function of the injection current at temperature $T/T_c = 0.9$ for $P_1 = 0.8$ and three values of $P_2 = 0$, 0.4, and 0.8. Other parameters are taken to be $R_1 = R_2$ ($g_i = 1/2$) and $\Gamma_S = 0$. In the case that FM1 and FM2 are the same ferromagnets (left panel), the critical current $J_c$ in the A alignment steeply decreases and vanishes at a small value of $I_{inj}$, whereas $J_c$ in the F alignment shows no dependence on $I_{inj}$. In the case that FM1 and FM2 are different (middle panel), the critical current decreases with increasing injection current in both alignments but in different way; $J_c$ decreases more slowly in the F alignment than in the A alignment. If one of the ferromagnets, FM2, is replaced by a normal metal (N), we have a heterostructure junction FM1/SC/N, which corresponds to the junction with $P_2 = 0$ (right panel). The calculated result for $P_2 = 0$ explains the critical current suppression by spin injection observed in the heterostructure junctions consisting of a high-$T_c$ SC and a ferromagnetic manganite with $P \sim 100\%$ [81-85].

# 4.5 Other quantum effects on TMR

## 4.5.1 Magnetic quantum point contact

The investigation of quantum effects on the mechanical and electrical properties of nanostructures has been attracted much interest. Recently, attention has focused on the conductance quantization in metallic nanowires, where an atomic scale constriction called point contact (PC) is made by pulling off two electrodes in contact [93-96]. The quantized conductances of integer multiples of $e^2/h$ in non-magnetic PC is well explained by Landauer's formula [97] of quantum ballistic transport combined with the adiabatic principle [98, 99]: for a non-magnetic PC, the spin-up and spin-down electrons make the same contribution and the unit of the conductance quantization is $2e^2/h$.

However, if the PC is made of a magnetic-metal such as Fe, Co and Ni, the exchange energy removes the spin degeneracy of conduction electrons and an atomic scale domain wall (DW) is created in the PC [100]. The conductance depends on the relative orientation of magnetizations between left and right electrodes, like magnetic tunnel junctions [4, 5, 101]. The spin dependent transport such as the TMR in magnetic tunnel junctions and GMR in magnetic multilayers [1] is of current interest both in fundamental physics and application to spin-electronics. It is then intriguing to ask how the exchange energy and the DW affect the electron transport in a magnetic PC. Recently, the following fascinating experimental results have been reported in Ni PC: the MR in excess of 200% [102, 103], the spin-dependent conductance quantization [104] and the $2e^2/h$ to $e^2/h$ switching of the quantized conductance [105].

In this subsection, we study the electron transport through a magnetic PC by using the recursion-transfer-matrix (RTM) method [106, 107, 108]. When a magnetic field is applied, the magnetizations of left and right electrodes are aligned parallel. In this ferromagnetic (F) alignment, the quantized conductance of odd integer multiples of $e^2/h$ appears, since the spin-up and spin-down electrons contribute to the conductance in a different way. On the other hand, in the absence of the magnetic field, the system is in the antiferromagnetic (A) alignment, where the magnetizations of left and right electrodes are antiparallel. The DW is created inside the constriction. We show that the spin precession of conduction electrons is forbidden in such an atomic scale DW. The contributions to the conductance from the spin-up and spin-down electrons are the same and the unit of the conductance quantization is $2e^2/h$. We also show that the MR is strongly enhanced for the narrow PC and oscillates with the conductance. Our study explains recent experimental results [103, 104, 105], and provides a new direction for spin-electronic devices with an atomic scale DW.

#### 4.5.1.1 Conductance quantization

An intuitive understanding of the physics of conductance quantization is given by the Landauer's formula [97] combined with an adiabatic approximation [98, 99]. Let us consider the non-magnetic PC connected to the left and right reservoirs as shown in Fig. 4.42(a). The cross-sections perpendicular to the contact axis are circles of radius $a(z)$, which take the minimum value $a_0$ at the center of the contact. If we consider the leads of straight wires with constant width connecting the reservoirs to the contact, there is a well defined transverse modes called "*channel*" in the leads. In a free electron model, each channel can be labeled by the set of quantum numbers in the transverse direction. By using the generalized Landauer's formula developed by Büttiker *et al.* [97], the conductance $G$ of the PC is given by

$$G = \frac{2e^2}{h} \sum_{ij} |t_{ij}|^2, \tag{4.178}$$

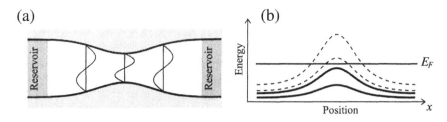

Figure 4.42: (a) Schematic representation of a non-magnetic PC. (b) The effective potential barrier $\varepsilon_{mn}(x)$ corresponding to each adiabatic transverse mode $(m, n)$. The finite number of modes with a barrier less than $E_F$ contributes each with the conductance quantum $2e^2/h$ for non-magnetic PC. The solid (dotted) lines indicate the transmitting (reflecting) channels.

where $e$ is the electron charge, $h$ is Planck's constant and the matrix element $t_{ij}$ is the transmission amplitude for an electron with incoming channel $j$ being scattered elastically into outgoing channel $i$. The prefactor 2 in the right hand side of Eq. (4.178) is due to the spin degeneracy of the amplitude $t_{ij}$. The transmission amplitude $t_{ij}$ is obtained by solving the 3D Schrödinger equation with the scattering boundary condition. Assuming that the wave function vanishes at the boundary of the geometrical constriction, the 3D wave function $\psi(x, y, z)$ is expanded using the energy eigenstate of the transverse direction $y$ and $z$ at each cross-section $x$ as,

$$\psi(x, y, z) = \sum_{mn} J_m \left( \frac{\gamma_{mn} r}{a(x)} \right) e^{im\varphi} \phi_{mn}(x), \qquad (4.179)$$

where $y = r \cos(\varphi)$, $z = r \sin(\varphi)$, $\gamma_{mn}$ is the $n$th zero of the Bessel function $J_m(\gamma_{mn} r/a(x))$ and $\phi_{mn}(x)$ are the $x$-dependent expansion coefficients [109]. The channel is labeled by the set of quantum numbers $(m, n)$. Substituting Eq. (4.179) to the Schrödinger equation and neglecting the coupling among different channels, we have

$$\left( -\frac{\hbar^2}{2m} \frac{\partial^2}{\partial x^2} + \varepsilon_{mn}(x) \right) \phi_{mn}(x) = E_F \phi_{mn}(x), \qquad (4.180)$$

where

$$\varepsilon_{mn}(x) = \frac{\hbar^2}{2m} \frac{\gamma_{mn}^2}{a(x)} \qquad (4.181)$$

is the effective potential barrier for $\phi_{mn}(x)$. In the strictly adiabatic case the effective potential barrier $\varepsilon_{mn}(x)$ either transmit or reflect an electron. Only the finite number of transverse modes with $\varepsilon_{mn}(x) < E_F$ all the way through the contact will be transmitted as shown in Fig. 4.42(b). Hence, the conductance given by Eq. (4.178) is written as

$$G = \frac{2e^2}{h} \sum_{mn} \theta \left( E_F - \varepsilon_{mn}(x) \right) = \frac{2e^2}{h} N_t, \qquad (4.182)$$

where $N_t$ is the number of transmitting channels. Since the Fermi energy is expressed as $E_F = \hbar^2 k_F^2/2m$, the number $N_t$ is given by the number of the set of quantum numbers $(m, n)$ satisfying the condition $\gamma_{mn} < k_F a_0$, where $a_0$ is the minimum value of the radius $a(x)$ in the contact. By narrowing the contact we can reduce the number of transmitting channels, and thus the conductance.

### 4.5.1.2    Model and method

Landauer's formula and the adiabatic picture give the qualitative understanding of the conductance quantization. However, for the quantitative discussion

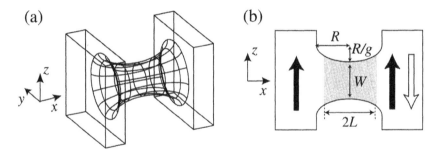

Figure 4.43: (a) Geometry of the constriction potential $V(r_\parallel, x)$. (b) Cross-section of the constriction potential in the $xz$ plane. Arrows represent the magnetization vectors of electrodes. In the right electrode, the filled (hollow) arrow represents the magnetization vector for the F (A) alignment. The shaded region in the constriction represents the DW with thickness $2L$ in the A alignment.

in the real system, the mixing of the channels can not be neglected and we have to solve the 3D Schrödinger equation directly. In this subsection, we introduce the model and numerical method to solve scattering problem in the 3D magnetic PC.

We consider the system consisting of two flat electrodes connected by a contact as shown in Figs. 4.43(a) and (b). The magnetization vectors of the left and right electrodes are indicated by the arrows in Fig. 4.43(b). For the A alignment, the atomic scale DW indicated by shading is created in the contact. The effective one-electron Hamiltonian is given by,

$$\mathcal{H} = -\frac{\hbar^2}{2m}\nabla^2 + V(r_\parallel, x) - \boldsymbol{h}(x) \cdot \boldsymbol{\sigma}, \qquad (4.183)$$

where $V(r_\parallel, x)$ is the constriction potential, $\boldsymbol{h}(x)$ is the exchange field and $\boldsymbol{\sigma}$ is the Pauli spin matrix. Here, $r_\parallel = (y, z)$ is the position vector in the $yz$ plane. We define the constriction potential $V(r_\parallel, x)$ as [107],

$$V(r_\parallel, x) = V_0 \begin{cases} \theta(R + x)\theta(R - x), & r_\parallel \geq \frac{R}{g} + \frac{W}{2}, \\ \theta(R - \sqrt{x^2 + (R + gW/2 - gr_\parallel)^2}, & r_\parallel < \frac{R}{g} + \frac{W}{2}, \end{cases} \qquad (4.184)$$

where $V_0$ is the height of the potential, $R$ is the radius of the elliptic envelope with a deformation parameter $g$ and $W$ is the width of the constriction. We take the $z$ axis to be parallel to the exchange field in the left electrode. For the F alignment, the exchange field is constant $\boldsymbol{h}(x) = (0, 0, h_0)$. For the A alignment, on the contrary, $\boldsymbol{h}(x)$ is not constant inside the DW. The detailed form of the exchange field $\boldsymbol{h}(x)$ for both classical and quantum DW will be given later.

In order to obtain the stationary scattering states, we employ the RTM method [106] including the spin degree of freedom. A 2D supercell structure is considered in the $y$ and $z$ directions. We take a unit of the supercell large enough to regard the transmission through the supercell as that in the non-periodic potential. Owing to the Bloch's theorem, electronic states are written in terms of the discrete reciprocal lattice vectors $\boldsymbol{K}_\parallel^n$ in the $y$ and $z$ directions. The $m$th stationary scattering state with spin $\sigma$ is written as,

$$\psi_{m\sigma}(\boldsymbol{r}_\parallel, x) = e^{i\boldsymbol{k}_\parallel \cdot \boldsymbol{r}_\parallel} \sum_{n\sigma'} \phi_{n\sigma',m\sigma}(x) e^{i\boldsymbol{K}_\parallel^n \cdot \boldsymbol{r}_\parallel}, \qquad (4.185)$$

where $\boldsymbol{k}_\parallel$ is the conserved Bloch $k$-vector and $\phi_{n\sigma',m\sigma}(x)$ are unknown coefficients to be solved. The combination of the index for the reciprocal lattice vector and the spin, $(n, \sigma)$, defines the "channel". The number of channels $N_c$ is truncated by including only the set of $\boldsymbol{K}_\parallel$ satisfying $|\boldsymbol{k}_\parallel + \boldsymbol{K}_\parallel^n|^2 < (2mE_c)/\hbar^2$, where $E_c$ is the cutoff energy.

Outside the left (right) boundary of the scattering region, electrons with spin $\sigma = \uparrow, \downarrow$ feel the constant potential, $U_L^\downarrow = U_R^\uparrow = h_0$, $U_L^\uparrow = U_R^\downarrow = -h_0$ and $\psi_{m\sigma}(\boldsymbol{r})$ is written as,

$$\psi_{m\sigma}(\boldsymbol{r}) = \begin{cases} e^{i\theta_{m\sigma}} e^{ik_{m\sigma}^L x} e^{i(\boldsymbol{k}_\parallel + \boldsymbol{K}_\parallel^m) \cdot \boldsymbol{r}_\parallel} \\ \quad + \sum_{n\sigma'} r_{n\sigma',m\sigma} e^{-ik_{n\sigma'}^L x} e^{i(\boldsymbol{k}_\parallel + \boldsymbol{K}_\parallel^n) \cdot \boldsymbol{r}_\parallel}, & x \leq x_L \\ \sum_{n\sigma'} t_{n\sigma',m\sigma} e^{-ik_{n\sigma'}^R x} e^{i(\boldsymbol{k}_\parallel + \boldsymbol{K}_\parallel^n) \cdot \boldsymbol{r}_\parallel}, & x \geq x_R, \end{cases} \qquad (4.186)$$

where $r_{n\sigma',m\sigma}$ ($t_{n\sigma',m\sigma}$) is the reflection (transmission) coefficient, $\theta_{m\sigma}$ is the initial phase, $k_{n\sigma'}^{L(R)} = \{(2m/\hbar^2)(E_F - U_{L(R)}^{\sigma'}) - |\boldsymbol{k}_\parallel + \boldsymbol{K}_\parallel^n|^2\}^{1/2}$ and $x_{L(R)}$ is the left(right) boundary of the scattering region. The coefficients $r_{n\sigma',m\sigma}$ and $t_{n\sigma',m\sigma}$ are obtained by solving the RTM equation. The transmission matrix of the system, $\mathsf{T}$, is expressed in terms of the coefficients $t_{n\sigma',m\sigma}$ as

$$\mathsf{T} = (\mathsf{k}_R^{\frac{1}{2}})^T \, \mathsf{t} \, \mathsf{k}_L^{-\frac{1}{2}}, \qquad (4.187)$$

where $\mathsf{k}_R^{\frac{1}{2}}(\mathsf{k}_L^{-\frac{1}{2}})$ is a $N \times M$ rectangular matrix whose $i, j$ elements are given by $k_i^{R\frac{1}{2}}\delta_{i,j}(k_i^{L-\frac{1}{2}}\delta_{i,j})$ and $M$ is the number of the open channels deep in the right (left) electrode. The reflection matrix, $\mathsf{R}$, is expressed in terms of the coefficients $r_{n\sigma',m\sigma}$ as

$$\mathsf{R} = (\mathsf{k}_L^{\frac{1}{2}})^T \, \mathsf{r} \, \mathsf{k}_R^{-\frac{1}{2}}. \qquad (4.188)$$

The unitary relation of the scattering matrices $\mathsf{T}$ and $\mathsf{R}$:

$$\mathsf{T}^\dagger \mathsf{T} + \mathsf{R}^\dagger \mathsf{R} = \mathsf{I} \qquad (4.189)$$

guarantees the current conservation. From Eq. (4.186), the conductance per supercell is calculated as

$$G = \frac{e^2}{h} \int dk_{\parallel} \frac{S}{(2\pi)^2} \mathrm{tr}\left(\mathsf{T}^\dagger \mathsf{T}\right). \tag{4.190}$$

For sufficiently large $yz$ unit cell, it is sufficient to use only $\Gamma$ point $k_{\parallel} = 0$ and the conductance is written without an integration as [107, 108],

$$G = \frac{e^2}{h} \mathrm{tr}\left(\mathsf{T}^\dagger \mathsf{T}\right), \tag{4.191}$$

which coincides with Landauer's formula [97].

We choose the commonly accepted values of the material parameters for typical ferromagnetic metals of Ni, Co and Fe [110, 111]. The Fermi energy and the height of the constriction potential are taken to be $E_F = 3.8\,\mathrm{eV}$ ($k_F = 1.0\,\text{Å}^{-1}$) and $V_0 = 2E_F$, respectively. The length of the constriction and the thickness of the DW are taken to be $2R = 2L = 10\,\text{Å}$ corresponding to $3 \sim 5$ atoms and the deformation parameter is $g = 10$. We choose the magnitude of the exchange field $h_0 = 0.3 \sim 0.7\,\mathrm{eV}$ and replace the step function $\theta$ in Eq. (4.184) by the Fermi function with a width of $0.25\,\text{Å}$ to make the constriction potential smooth. We use the square supercell with linear dimension of $20\,\text{Å}$ in the $y$ and $z$ directions. The left(right) boundary of scattering region is taken to be $x_{L(R)} = -10\,\text{Å}$ ($10\,\text{Å}$) and the size of the mesh is $0.2\,\text{Å}$ in the $x$ direction. The cutoff energy is $E_c = 21.8\,\mathrm{eV}$ and 354 channels are used in the numerical calculation.

### 4.5.1.3 Spin dependent conductance quantization

The conductance curves for a non-magnetic PC, i.e., $h(z) = 0$, are shown as a function of the width of the constriction, $W$, in Fig. 4.44(a). The conductances for spin-up and spin-down electrons, which are indicated by the dotted line, are degenerate. Therefore, the conductance never takes the quantized value of an odd integer multiples of $e^2/h$. The total conductance represented by the solid line has plateaus at $G = 2, 6, 10, 12 \times (e^2/h)$. The missing of the plateau at $4e^2/h$ and $8e^2/h$ is due to the rotational symmetry of the constriction potential. In the adiabatic picture, the channel becomes transmitting at plateaus of $G/(e^2/h) = 2, 6, 10, 12$ correspond to the quantum number $(m, n) = (0, 1), (\pm 1, 1), (\pm 2, 1), (0, 2)$.

For a magnetic PC, the degeneracy of the spin-up and spin-down conductances is removed by the exchange energy and plateaus of odd integer multiples of $e^2/h$ appear for the F alignment as shown in Fig. 4.44(b). The dotted and dot-dashed lines indicate the conductances for spin-up and spin-down electrons, respectively. In the adiabatic picture, the number of transmission channels are determined by the condition that $\varepsilon_{mn}(x) - \sigma h_0 < E_F$. If we rewrite

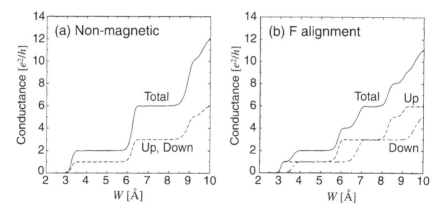

Figure 4.44: (a) The conductance curves for the non-magnetic PC ($h_0 = 0$). The conductances for the spin-up and spin-down electrons are degenerate and plotted by the dotted line. The total conductance is plotted by the solid line. (b) The conductance curves in the F alignment of the magnetic PC. The exchange field is $h_0 = 0.5\,\mathrm{eV}$. The conductance for the spin-up (spin-down) electrons is plotted by the dashed (dot-dashed) line. The total conductance is plotted by the solid line.

the condition as $\varepsilon_{mn}(x) < (E_F + \sigma h_0)$, the number of transmission channels for spin-up (spin-down) electrons corresponds to that for non-magnetic PC with the larger (smaller) Fermi energy. For spin-up (spin-down) electrons, the width of the PC, $W$, where the number of transmission channels changes, becomes smaller (larger) than that for a non-magnetic PC. As shown in Fig. 4.44(b), the total conductance shows plateaus at $G = 1, 2, 4, 9 \times (e^2/h)$. The width $W$ at which the new transmitting channel opens for spin-up (spin-down) electrons decreases (increases) as the exchange filed increases. This kind of the spin-dependent conductance quantization was first observed in semiconductor PCs under high magnetic fields [112, 113], where the spin degeneracy is removed by the Zeeman energy.

### 4.5.1.4   Effect of the classical domain wall

Let us move on to the effect of the atomic scale DW for the A alignment. Bruno [100] has studied such a geometrically constrained DW, and pointed out that when the cross-section of the constriction is much smaller than that of the electrode, the width of the geometrically constrained DW is essentially given by the length of the constriction, which can be considerably smaller than the width of a Bloch or Néel wall for the PCs made of magnetic-metals. In the classical theory of the geometrically constrained DW, the exchange field $\boldsymbol{h}(x)$

is given by [100, 110]

$$h_x(x) = 0, \quad h_y(x) = h_0\theta(L - |x|)\cos(Qx),$$
$$h_z(x) = h_0\{\theta(-L - x) - \theta(-L + x) - \theta(L - |x|)\sin(Qx)\}, \quad (4.192)$$

where $L$ is half the thickness of the DW and $Q = \pi/2L$.

Before studying the conductance quantization, we show how the spin of a conduction electron rotates without the constriction potential. We consider the following 1D Hamiltonian:

$$H = -\frac{\hbar^2}{2m}\frac{\partial^2}{\partial x^2} - \boldsymbol{h}(x) \cdot \boldsymbol{\sigma}, \quad (4.193)$$

where $\boldsymbol{h}(x)$ is given by Eq. (4.192). The analytical results of the probability current density is obtained by connecting wave functions at the boundary of the DW. For the analytical calculation, it is convenient to rotate the spin-quantization axis parallel to the $x$ axis. In this case, the eigenstates of the Hamiltonian (Eq. (4.193)) in the DW region are the same as those of a "spin spiral" [114, 115]. Assuming that the spin-up electron ($s_x = 1/2$) is incident on the left electrode as shown in Fig. 4.45(a), the wave functions in the left (L) electrode, DW, and right (R) electrode are given by,

$$\psi_L = \begin{pmatrix} e^{ik_L^\uparrow x} \\ e^{ik_L^\uparrow x} \end{pmatrix} + r_{\uparrow\uparrow}\begin{pmatrix} e^{-ik_L^\uparrow x} \\ e^{-ik_L^\uparrow x} \end{pmatrix} + r_{\downarrow\uparrow}\begin{pmatrix} e^{-ik_L^\downarrow x} \\ -e^{-ik_L^\downarrow x} \end{pmatrix},$$

$$\psi_{DW} = \sum_{j=1}^{4} D_j\begin{pmatrix} e^{ik_D^j x} \\ \alpha_j e^{i(k_D^j + Q)x} \end{pmatrix},$$

$$\psi_R = t_{\uparrow\uparrow}\begin{pmatrix} e^{ik_R^\uparrow x} \\ e^{ik_R^\uparrow x} \end{pmatrix} + t_{\downarrow\uparrow}\begin{pmatrix} e^{ik_R^\downarrow x} \\ -e^{ik_R^\downarrow x} \end{pmatrix}, \quad (4.194)$$

where $k_L^\uparrow = k_R^\downarrow = \sqrt{2m(E_F - h_0)}/\hbar$, $k_L^\downarrow = k_R^\uparrow = \sqrt{2m(E_F + h_0)}/\hbar$, $k_D^j$ are the four solutions of the equation $(\varepsilon_{k_D^j} - E_F)(\varepsilon_{k_D^j + Q} - E_F) = h_0^2$ and the coefficients $\alpha_j = (E_F - \varepsilon_{k_D^j})/ih_0$ with $\varepsilon_k = \hbar^2 k^2/2m$.

The spin precession is often studied in analogy with the magnetic resonance. The dot-dashed line represents the probability that the spin $s_x = \hbar/2$ at $t = 0$ will have the value $s_x = -\hbar/2$ at $t = 2L/\bar{v}_F$ in a magnetic filed $h_0$ rotating with a frequency $\pi\bar{v}_F/2L$, where $\bar{v}_F = [\sqrt{2m(E_F + h_0)} + \sqrt{2m(E_F - h_0)}]/2$ is the mean Fermi velocity. Let us introduce the spin precession length $l_s \equiv \hbar\bar{v}_F/4h_0$. For the exchange field $h_0 = 0.5\,\text{eV}$, $l_s$ is estimated to be 24 Å. The spin flip probability is given by $\xi^2/(1 + \xi^2)$ with $\xi = 2L/l_s$.

In Fig. 4.45, the transmission probability, $T_{\downarrow\uparrow(\uparrow\uparrow)} = |t_{\downarrow\uparrow(\uparrow\uparrow)}|^2 k_R^{\downarrow(\uparrow)}/k_L^\uparrow$, for the exchange energy $h_0 = 0.5\,\text{eV}$ is plotted by the solid (dotted) line. The probability $T_{\downarrow\uparrow}$ oscillates between a lower envelope given by $\xi^2/(1 + \xi^2)$ and

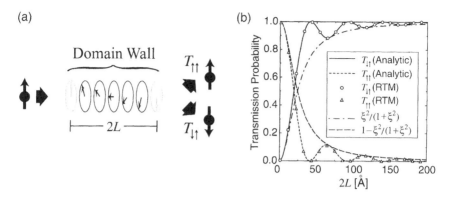

Figure 4.45: (a) Magnetization vector of 1D classical DW with thickness of $2L$. The arrow with a filled circle in the left side indicates the spin of incident electron. The arrows with filled circles in the right side indicate the spins of outgoing electrons. The transmission probabilities for the incident electron with spin-up are shown in the panel (b). The analytical results of the probability current densities $T_{\downarrow\uparrow}$ and $T_{\uparrow\uparrow}$ are plotted by the solid and dotted lines, respectively. The circles (triangles) indicate $T_{\downarrow\uparrow}$ ($T_{\uparrow\uparrow}$) obtained numerically by using the RTM method. The dashed (dot-dashed) line is the spin conservation (flip) probability in the magnetic filed oscillating as a function of time.

unity, while $T_{\uparrow\uparrow}$ oscillates between zero and a upper envelope $1 - \xi^2/(1 + \xi^2)$. The lower and upper envelopes represent the spin-flip and spin-conservation probabilities in an oscillating magnetic field. The probability current density for the reflection, $|r_{\uparrow\uparrow}|^2 k_L^\uparrow/k_L^\uparrow + |r_{\downarrow\uparrow}|^2 k_L^\downarrow/k_L^\uparrow$, is less than 0.005 even for the 1-Å thick DW and decreases as the thickness of the DW increases. We also plot the numerical result of $T_{\downarrow\uparrow}$ and $T_{\uparrow\uparrow}$ obtained by the RTM method in Fig. 4.45(a). The excellent agreement between the analytic and numerical results indicates that the RTM method can accurately describe the spin precession of conduction electrons.

As shown in Fig. 4.45(b), the transmission probability to the spin-down state, $T_{\downarrow\uparrow}$, is as small as 0.16 at $2L = 10$ Å. This means that the spin of the conduction electron hardly rotates through the 1D classical DW of the size of the PCs that show the conductance quantization. In the PC, however, the geometrical constriction plays a crucial role in the spin precession through the classical DW. In the adiabatic picture, the velocity in the $x$ direction for channel $n$ is given by $v_x^n = \sqrt{(2/m)(E_F - E_{n\|})}$, where $E_{n\|}$ is the energy eigenvalue of the transverse mode. $E_{n\|}$ is a decreasing function of $W$ and the channel $n$ opens when $E_{n\|}$ becomes smaller than $E_F$. Therefore, the velocity of electrons transmitting through the channel $n$ becomes very small: $v_x^n \ll v_F$, where $v_F = \sqrt{(2/m)E_F}$ is the Fermi velocity.

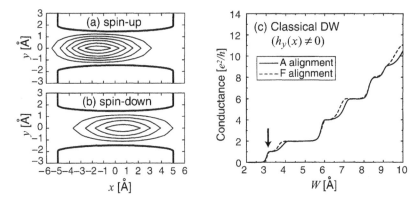

Figure 4.46: Panels (a) and (b) shows the contour plots of the spin-up and spin-down charge densities for the transmitting channel at $W = 3.2\,\text{Å}$ indicated by an arrow in the panel (c). (c) The conductance curve of the classical DW in the A alignment (solid line) and in the F alignment (dotted line). The spin-precession of conduction electron in the DW is allowed. The parameters are the same as those in Fig. 4.44.

In Figs. 4.46(a) and (b), we show the spin-up and spin-down charge densities of the transmitting channel [108, 116] for the A alignment at $W = 3.2\,\text{Å}$, respectively. Only one transmitting channel opens at $W = 3.2\,\text{Å}$, and the position is indicated by the arrow in Fig. 4.46(c). The incident spin-up electron tracks the local exchange field and transmits as the spin-down electron. Through the DW, electrons feel a nearly constant exchange field as if they were in the F alignment. Therefore, the conductance curve for the A alignment is similar to that for the F alignment as shown in Fig. 4.46(c). However, it has been experimentally observed that the sequence of the quantized conductances is different between the F and A alignments; the conductance plateaus of odd integer multiples of $e^2/h$ appear only for the F alignment [104, 105].

### 4.5.1.5 Effect of the quantum domain wall

This discrepancy can be resolved by considering an atomic scale DW on the basis of the quantum theory. As pointed out by Bruno [100], the DW is about the size of the PC. For such an atomic scale DW, we have to derive the exchange field $\boldsymbol{h}(x)$ on the basis of the quantum theory. The narrow band $d$-states are susceptible to disorder due to the small hopping matrix element and are easily localized [117]. Let us examine the DW in a PC by using the following Heisenberg ferromagnet [118, 119],

$$\mathcal{H}_{\text{DW}} = -J_{\text{DW}} \sum_{<i,j>} \boldsymbol{S}_i \cdot \boldsymbol{S}_j - \alpha \left( \sum_{i \in F_L} S_i^z - \sum_{i \in F_R} S_i^z \right), \qquad (4.195)$$

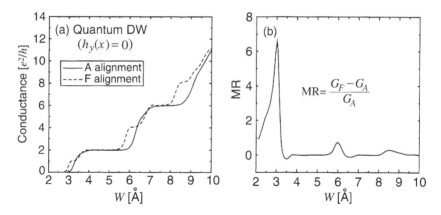

Figure 4.47: (a) The conductance curves of a magnetic PC in the F alignment (solid curve) and the A alignment (dashed curve). (b) Magnetoresistance of the magnetic PC as a function of the width of the constriction.

where $\boldsymbol{S}_i$ is the localized $3d$ spin at site $i$. We consider the nearest-neighbor interactions with coupling constant $J_{\mathrm{DW}}$ and neglect the anisotropy. The second term contains the interactions of spins in the left (right) electrode $F_{L(R)}$ with the coupling constant $\alpha$. We assume that $S_i = 1$ for Ni PC. The eigenstates of the quantum DW given by Eq. (4.195) are labeled by the $z$ component of the total spin $S^z \equiv \sum_i S_i^z$ and the ground state is $S^z = 0$. The exchange energy of the effective one-electron Hamiltonian $-\boldsymbol{h}(x) \cdot \boldsymbol{\sigma}$ is expressed as $-J \sum_i \boldsymbol{s} \cdot \langle \boldsymbol{S}_i \rangle$, where $\boldsymbol{s}$ is the spin of a conduction electron and $\langle \boldsymbol{S}_i \rangle$ represents the expectation value of the localized spin $\boldsymbol{S}_i$ and $J$ is the corresponding coupling constant [120]. Since the DW consists of a few atoms and is strongly pinned by the geometrical constriction, the DW has a large excitation energy and the expectation value $\langle \boldsymbol{S}_i \rangle$ is evaluated in the ground state with $S^z = 0$. We studied the $N \leq 8$ site 1D DW with $S = 1$ by numerically diagonalizing the Hamiltonian $\mathcal{H}_{\mathrm{DW}}$. We find that the $\langle S_i^z \rangle$ is well fitted by $h_z(x)$ in Eq. (4.192) and the excitation energy for $S^z = \pm 1$ is large. For example, the excitation energy is $1.1\, J_{\mathrm{DW}}$ for $N = 4$ and $J_{\mathrm{DW}} = \alpha$. One crucial property of such an atomic scale quantum DW is that the expectation values $\langle S_i^x \rangle = \langle S_i^y \rangle = 0$ for all sites $i$, because the operators $S_i^x$ and $S_i^y$ have only nonzero matrix elements between states of different eigenvalue of $S_z$ [118]. Therefore, the spin-mixing term vanishes, i.e., $-h_y(x)\sigma_y = -J \sum_i \boldsymbol{s} \cdot \langle S_i^y \rangle = 0$ in Eq. (4.183) for the atomic scale quantum DW. The spin of the conduction electrons cannot rotate in the DW.

The conductance curve for the A alignment with a quantum DW are plotted in Fig. 4.47(b). The sequence of the quantized conductances is clearly different between F and A alignments. In the adiabatic picture, the number of transmitting channels is the same for spin-up and spin-down electrons because the exchange energy $-h_z(x)\sigma_z$ is an odd function of $x$. The sequence of the quantized conductances is the same as that for a non-magnetic PC shown in

Fig. 4.44(a). The conductance curve shows clear plateaus at $G = 2e^2/h$ and $6e^2/h$ and the plateaus at odd integer multiples of $e^2/h$ disappear. Comparing Figs. 4.46(c) and 4.47(a), we conclude that the recent experimental results that the sequence of the quantized conductances is different between A and F alignments [104, 105] is the direct consequence of the fact that the spin of conduction electrons cannot rotate in the atomic scale quantum DW.

### 4.5.1.6 Magnetoresistance

Figure 4.47(b) shows the MR calculated by the formula: MR=$(G_F/G_A) - 1$, where $G_{F(A)}$ is the total conductance for the F (A) alignment given in Fig. 4.47(a). A strong enhancement of the MR appears at $W \simeq 3$ Å, where the first transmitting channel opens for the F alignment, but no transmitting channels for the A alignment. The key point is that the DW makes the number of transmitting channels different between F and A alignments as shown in Fig. 4.47(a). Note that the conductance plateau at integer multiples of $e^2/h$ means that the scattering intrinsic to the DW [121, 115] is negligible. In addition, the MR oscillates with $W$ as shown in Fig. 4.47(b). Since the difference in the conductances (or the number of transmitting channels) between the A and F alignments does not increase with $W$ as shown in Fig. 4.47(a), the magnitude of the oscillation decreases with $W$.

The MR increases as the exchange field $h_0$ increases. Figures 4.48(a)-(c) shows the MR for the exchange energy $h_0 = 0.3$, 0.5 and 0.7 eV, respectively.

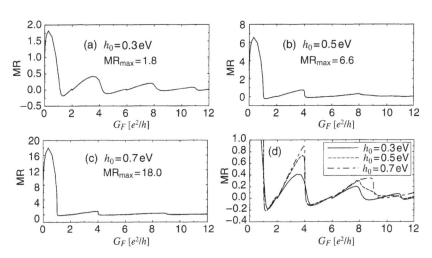

Figure 4.48: Magnetoresistances (MR) as a function of the conductance $G_F$ in the F alignment. Panels (a), (b) and (c) show the MR for different values of exchange fields $h_0 = 0.3$, 0.5 and 0.7 eV, respectively. The enlarged view of MR is shown in panel (d).

The horizontal axis is taken to be the conductance for the F alignment $G_F$ in order to compare the present result with experimental ones in Refs. [102, 103]. The MR shows a strong enhancement at $G_F/(e^2/h) \simeq 1$ ($W \simeq 3$ Å). The maximum values of MR, $\mathrm{MR_{max}}$, for the exchange fields $h_0 = 0.3$, 0.5 and 0.7 eV are 1.8, 6.6 and 18.0, respectively. The enhancement of MR is expected to be very large if magnetic-metals with large exchange field such as Co and Fe are used. We also find the MR oscillates with $G_F$ as shown in Fig. 4.48(d). Our results explain the experimentally observed MR enhancement and oscillation [102, 103].

### 4.5.2  Andreev reflection

A new technique based on Andreev reflection between a point contact and sample is used to determine the spin polarization of a ferromagnet [122, 123]. Andreev reflection is a basic process that occurs at the interface between a normal metal and a superconductor [124]. In the normal side of the interface, an incident electron with spin $\sigma$ takes another electron with opposite spin $-\sigma$ to enter the superconductor in the form of a Copper pair, thereby reflecting a positively charged hole with a reversed spin direction. This process allows supercurrent to flow across the N/SC interface for low bias voltages below the superconducting gap $\Delta$.

The Andreev reflection at the interface of a ferromagnetic metal and a superconductor is strongly modified because of the incident electrons and the reflected holes occupy the states of the opposite spin-bands in the ferromagnet. de Jong and Beenakker [125] presents an intuitive and simple description of the conductance through a ballistic ferromagnet/superconductor point contact at zero temperature. A ferromagnet is contacted through a small area with a superconductor, in which the number of up-spin channels ($N_\uparrow$) is larger than that of down-spin channels ($N_\downarrow$), i.e., $N_\uparrow \geq N_\downarrow$. When the superconductor is in the normal state, all scattering channels (transverse modes in the point contact at the Fermi level) are fully transmitted, yielding the conductance

$$G_{FN} = \frac{e^2}{\hbar} (N_\uparrow + N_\downarrow).$$  (4.196)

In the superconducting state, the spin-down electrons of all the $N_\downarrow$ channels are Andreev reflected into spin-up holes. They give a double contribution to the conductance since $2e$ is transferred at each Andreev reflection. However, only a fraction ($N_\downarrow/N_\uparrow$) of the $N_\uparrow$ channels can be Andreev reflected, because the density of states in the spin-down band is smaller than in the spin-up band. Therefore, the resulting conductance at zero bias ($V = 0$) is

$$G_{FS} = \frac{e^2}{\hbar} \left( 2N_\downarrow + 2\frac{N_\downarrow}{N_\downarrow}N_\uparrow \right) = 4N_\downarrow.$$  (4.197)

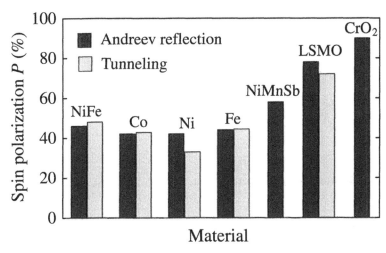

Figure 4.49: Comparison of the measured values of $P$ from two techniques: tunneling and Andreev reflection [122]. The tunneling date are the same as those in Table 4.1.

Taking the ratio of Eq. (4.197) to Eq. (4.196), we obtain the normalized conductance

$$\frac{G_{FS}}{G_{FN}} = 2\left(1 - P_c\right), \qquad (4.198)$$

where $P_c$ is the spin-polarization defined by

$$P_c = \frac{N_\uparrow - N_\downarrow}{N_\uparrow + N_\downarrow} = \frac{\mathcal{D}_\uparrow v_{F\uparrow} - \mathcal{D}_\downarrow v_{F\downarrow}}{\mathcal{D}_\uparrow v_{F\uparrow} + \mathcal{D}_\downarrow v_{F\downarrow}}. \qquad (4.199)$$

Here, we used $N_\sigma = k_{F\sigma}^2 A/4\pi$, $k_{F\downarrow}$ is the Fermi momentum of spin-$\sigma$ subband, and $A$ is the area of the contact.

The experimental results of $P_c$ by performing Andreev reflection with a Nb point contact are shown in Fig. 4.49 [122]. The measured samples are permalloy ($Ni_x Fe_{1-x}$), Ni, Co, Fe, NiMnSb, $La_{0.7}Sr_{0.3}MnO_3$, and $CrO_2$. The spin polarizations of various ferromagnets $P_c$ can be estimated directly from the conductance values at $V = 0$ using Eq. (4.198). It has been shown that many measurements are successfully carried out with other superconductors (NbN, Ta, and V). Even if the role of point contact and base electrode is reversed, a significant difference is not seen in the measured $G(V)$ curves or in the inferred value of $P_c$. The results obtained by Andreev reflection are compared with those obtained by tunneling experiments (see also Table 4.1) in Fig. 4.49. The agreement between the two methods is generally good.

# References

[1] M. N. Baibich, J. M. Broto, A. Fert, F. Nguyen Van Dau, F. Petroff, P. Etienne, G. Creuzet, A. Friederich and J. Chazelas, Phys. Rev. Lett. **61**, 2472 (1988).

[2] I. Giaever, Phys. Rev. Lett. **5**, 147 (1960); *ibid.*, **5**, 464 (1960).

[3] R. Meservey and P. M. Tedrow, Phys. Rep. **238**, 173 (1994).

[4] M. Julliere, Phys. Lett. **54A**, 225 (1975).

[5] S. Maekawa and U. Gäfvert, IEEE Trans. Magn. **18**, 707 (1982).

[6] T. Miyazaki and N. Tezuka, J. Magn. Magn. Mater. **139**, L231 (1995).

[7] J. S. Moodera, L. Kinder, T. M. Wong and R. Meservey, Phys. Rev. Lett. **74**, 3273 (1995).

[8] *Tunneling Phenomena in Solids*, edited by E. Burstein and S. Lundqvist (Plenum Press, New York, 1969).

[9] J. C. Slonczewski, Phys. Rev. B **39**, 6995 (1989).

[10] S. Zhang and P. Levy, Eur. Phys. J. B **10**, 599 (1999).

[11] J. S. Moodera and G. Mathon, J. Magn. Magn. Mater. **200**, 248 (1999).

[12] D. J. Monsma and S. S. P. Parkin, Appl. Phys. Lett. **77**, 720 (2000).

[13] D. C. Worledge and T. H. Geballe, Appl. Phys. Lett. **76**, 900 (2000).

[14] M. B. Stearns, J. Magn. Magn. Mater. **5**, 167 (1977).

[15] J. A. Hertz and K. Aoi, Phys. Rev. B **8**, 3252 (1973).

[16] D. Nyugen-Mahn, E. Y. Tsymbal, D. G. Pettifor, C. Arcangeli, R. Tank, O. K. Andersen and A. Pasturel, Mater. Res. Soc. Symp. Proc. **492**, 319 (1998).

[17] E. Y. Tsymbal and D. G. Pettifor, J. Appl. Phys. **87**, 5230 (2000).

[18] G. Prinz, J. Magn. Magn. Mater. **200**, 57 (1999).

[19] N. Tezuka and T. Miyazaki, J. Magn. Magn. Mater. **177-181**, 1283 (1998).

[20] J. G. Simmons, J. Appl. Phys. **34**, 1793 (1963).

[21] Y. Asano, A. Oguri and S. Maekawa, Phys. Rev. B **48**, 6192 (1993).

[22] J. Mathon, Phys. Rev. B **56**, 11810 (1997).

[23] E. Y. Tsymbal and D. G. Pettifor, Phys. Rev. B **58**, 432 (1998).

[24] J. Mathon and A. Umerski, Phys. Rev. B **60**, 1117 (1999).

[25] H. Itoh, A. Shibata, T. Kumazaki, J. Inoue and S. Maekawa, J. Phys. Soc. Jpn. **68**, 1632 (1999).

[26] J. M. MacLaren, X.-G. Zhang, W. H. Butler and X. Wang, Phys. Rev. B **59**, 5740 (1999).

[27] H. Itoh, T. Ohsawa and J. Inoue, Phys. Rev. Lett. **84**, 2501 (2000).

[28] D. J. Thouless and S. Kirkpatrick, J. Phys. C: Solid State Phys. **14**, 235 (1981).

[29] P. A. Lee and D. S. Fisher, Phys. Rev. Lett. **47**, 882 (1981).

[30] J. M. De Teresa, A. Barthélémy, A. Fert, J. P. Contour, R. Lyonnet, F. Montaigne, P. Seneor and A. Vaurès, Phys. Rev. Lett. **82**, 4288 (1999).

[31] S. Yuasa, T. Sato, E. Tamura, Y. Suzuki, K. A. H. Yamamoto and T. Katayama, Eur. Phys. Lett. **52**, 344 (2000).

[32] P. M. Levy and S. Zhang, Curr. Opin. Solid State Mater. Sci. **4**, 223 (1999).

[33] R. Wilkins, E. Ben-Jacob and R. C. Jaklevic, Phys. Rev. Lett. **63**, 801 (1989).

[34] N. S. Bakhvalov, G. S. Kazacha, K. K. Likharev and S. I. Serdyukova, Sov. Phys. JETP **68**, 581 (1989).

[35] *Single Charge Tunneling*, edited by H. Grabert and M. H. Devoret (Plenum Press, New York, 1992).

[36] K. K. Likharev, IEEE Trans. Magn. **23**, 1142 (1987).

[37] K. Mullen, E. Ben-Jacob, R. C. Jaklevic and Z. Schuss, Phys. Rev. B **37**, 98 (1988).

[38] B. Laikhtman, Phys. Rev. B **43**, 2731 (1991).

[39] J. C. Wan, K. A. McGreer, L. I. Glazman, A. M. Goldman and R. I. Shekhter, Phys. Rev. B **43**, 9381 (1991).

[40] J. Barnaś and A. Fert, Phys. Rev. Lett. **80**, 1058 (1998).

[41] K. Majumdar and S. Hershfield, Phys. Rev. B **57**, 11521 (1998).

[42] R. Wiesendanger, M. Bode and M. Getzlaff, Appl. Phys. Lett. **75**, 124 (1999).

[43] E. Bar-Sadeh, Y. Goldstein, C. Zhang, H. Deng, B. Abeles and O. Millo, Phys. Rev. B **50**, 8961 (1994).

[44] R. Desmicht, G. Faini, V. Cros, A. Fert, F. Petroff and A. Vaurès, Appl. Phys. Lett. **72**, 386 (1998).

[45] J. Chiba, K. Takanashi, S. Mitani and H. Fujimori, J. Magn. Soc. Jpn. **23**, 82 (1999).

[46] S. Hershfield, J. H. Davies, P. Hyldgaard, C. J. Stanton and J. W. Wilkins, Phys. Rev. B **47**, 1967 (1993).

[47] M. Eto, J. Phys. Soc. Jpn. **65**, 1523 (1996).

[48] J. A. Melsen, U. Hanke, H.-O. Müller and K.-A. Chao, Phys. Rev. B **55**, 10638 (1997).

[49] U. Volmar, U. Weber, R. Houbertz and U. Hartmann, Appl. Phys. A **66**, S735 (1998).

[50] S. Mitani, S. Takahashi, K. Takanashi, K. Yakushiji, S. Maekawa and H. Fujimori, Phys. Rev. Lett. **81**, 2799 (1998).

[51] J. Inoue and S. Maekawa, Phys. Rev. B **53**, R11927 (1996).

[52] T. A. Fulton and G. J. Dolan, Phys. Rev. Lett. **59**, 109 (1987).

[53] H. Devoret, D. Esteve, H. Grabert, G.-L. Ingold, H. Pothier and C. Urbina, Phys. Rev. Lett. **64**, 1824 (1990).

[54] H. Grabert, G.-L. Ingold, M. H. Devoret, D. Esteve, H. Pothier and C. Urbiba, Z. Phys B **84**, 143 (1991).

[55] D. V. Averin and Y. V. Nazarov, Phys. Rev. Lett. **65**, 2446 (1990).

[56] L. J. Geerligs, D. V. Averin and J. E. Mooij, Phys. Rev. Lett. **65**, 3037 (1990).

[57] P. Joyez, V. Bouchiat, D. Esteve, C. Urbina and M. H. Devoret, Phys. Rev. Lett. **79**, 1349 (1997).

[58] J. König, H. Schoeller and G. Schön, Phys. Rev. Lett. **78**, 4482 (1997).

[59] D. V. Averin and A. A. Odinstov, Phys. Lett. A **140**, 251 (1989).

[60] K. Ono, H. Shimada and Y. Ootuka, J. Phys. Soc. Jpn. **66**, 1261 (1997).

[61] Y. Ootuka, R. Matsuda, K. Ono and H. Shimada, Physica B **280**, 394 (2000).

[62] L. F. Schelp, A. Fert, F. Fettar, P. Holody, S. F. Lee, J. L. Maurice, F. Petroff and A. Vauré, Phys. Rev. B **56**, R5747 (1997).

[63] K. Inomata and Y. Saito, Appl. Phys. Lett. **73**, 1143 (1998).

[64] Y. Fukumoto, H. Kubota, Y. Ando and T. Miyazaki, Jpn. J. Appl. Phys. **38**, L932 (1999).

[65] S. Takahashi and S. Maekawa, Phys. Rev. Lett. **80**, 1758 (1998).

[66] D. V. Averin, Physica B **194-196**, 979 (1994).

[67] K. Ono, Ph.D. thesis, The University of Tokyo, 1998.

[68] H. Schoeller and G. Schön, Phys. Rev. B **50**, 18436 (1994).

[69] X. H. Wang and A. Brataas, Phys. Rev. Lett. **83**, 5138 (1999).

[70] B. Abeles, P. Sheng, M. D. Coutts and Y. Arei, Adv. Phys. **24**, 407 (1975).

[71] H. Fujimori, S. Mitani and S. Ohnuma, Mat. Sci. Eng. **B31**, 219 (1995).

[72] M. Ohnuma, K. Hono, E. Abe, H. Onodera, S. Mitani and H. Fujimori, J. Appl. Phys. **82**, 5646 (1997).

[73] Aronov, JETP Lett. **24**, 32 (1977).

[74] M. Johnson and R. H. Silsbee, Phys. Rev. Lett. **55**, 1790 (1985).

[75] M. Johnson and R. H. Silsbee, Phys. Rev. B **37**, 5326 (1988).

[76] T. Valet and A. Fert, Phys. Rev. B **48**, 7099 (1993).

[77] S. Hershfield and H. L. Zhao, Phys. Rev. B **56**, 3296 (1997).

[78] A. Fert and S. F. Lee, Phys. Rev. B **B53**, 6553 (1996).

[79] M. Johnson, Phys. Rev. Lett. **70**, 2142 (1993).

[80] M. Johnson, Appl. Phys. Lett. **65**, 1460 (1994).

[81] V. A. Vas'ko, V. A. Larkin, P. A. Kraus, K. R. Nikolaev, D. E. Grupp, C. A. Nordman and A. M. Goldman, Phys. Rev. Lett. **78**, 1134 (1997).

[82] Z. W. Dong, R. Ramesh, T. Venkatesan, M. Johnson, Z. Y. Chen, S. P. Pai, V. Talyansky, R. P. Sharma, R. Shreekala, C. J. Lobb and R. L. Greene, Appl. Phys. Lett. **71**, 1718 (1997).

[83] D. Koller, M. S. Osofsky, D. B. Chrisey, J. S. Horwitz, J. R. J. Soulen, R. M. Stroud, C. R. Eddy, J. Kim, R. C. Y. Auyeung, J. M. Byers, B. F. Woodfield, G. M. Daly, T. W. Clinton and M. Johnson, J. Appl. Phys. **83**, 6774 (1998).

[84] K. Lee, W. Wang, I. Iguchi, B. Friedman, T. Ishibashi and K. Sato, Appl. Phys. Lett. **75**, 1149 (1999).

[85] N.-C. Yeh, R. Vasquez, C. Fu, A. Samoilov, Y. Li and K. Vakili, Phys. Rev. B **60**, 10522 (1999).

[86] S. Takahashi, H. Imamura and S. Maekawa, Phys. Rev. Lett. **82**, 3911 (1999).

[87] S. Takahashi, H. Imamura and S. Maekawa, J. Appl. Phys. **87**, 5227 (2000).

[88] M. Tinkham, *Introduction to Superconductivity* (McGraw-Hill, New York, 1996).

[89] J. Clarke, Phys. Rev. Lett. **28**, 1363 (1972).

[90] C. Pethick and H. Smith, J. Phys. C **13**, 6313 (1980).

[91] S. Takahashi, H. Imamura and S. Maekawa, Physica C **341-348**, 1515 (2000).

[92] M. Tinkham, Phys. Rev. B **6**, 1747 (1972).

[93] J. M. Krans and J. M. van Ruitenbeek, Phys. Rev. B **50**, 17659 (1994).

[94] L. Olesen, E. Læsgaard, I. Stensgaard, F. Besenbacher, J. Schiøtz, P. Stoltze, K. W. Jacobsen and J. K. Nørskov, Phys. Rev. Lett. **72**, 2251 (1994).

[95] G. Rubio, N. Agraït and S. Vieira, Phys. Rev. Lett. **76**, 2302 (1996).

[96] J. L. Costa-Krämer, Phys. Rev. B **55**, R4875 (1997).

[97]  M. Büttiker, Y. Imry, R. Landauer and S. Pinhas, Phys. Rev. B **31**, 6207 (1985).

[98]  L. I. Glazman, G. B. Lesovik, D. E. Khmel'nitskii and I. Shekhter, JEPT Lett. **48**, 238 (1988).

[99]  A. Yacoby and Y. Imry, Phys. Rev. B **41**, 5341 (1990).

[100]  P. Bruno, Phys. Rev. Lett. **83**, 2425 (1999).

[101]  H. Imamura, S. Takahashi and S. Maekawa, Phys. Rev. B **59**, 6017 (1999).

[102]  G. Tatara, Y. W. Zhao, M. Muñoz and N. García, Phys. Rev. Lett. **83**, 2030 (1999).

[103]  N. García, M. Muñoz and Y.-W. Zhao, Phys. Rev. Lett. **82**, 2923 (1999).

[104]  H. Oshima and K. Miyano, Appl. Phys. Lett. **73**, 2203 (1998).

[105]  T. Ono, Y. Ooka, H. Miyajima and Y. Otani, Appl. Phys. Lett. **75**, 1622 (1999).

[106]  K. Hirose and M. Tsukada, Phys. Rev. B **51**, 5278 (1995).

[107]  M. Brandbyge, K. W. Jabsen and J. K. Nørskov, Phys. Rev. B **55**, 2637 (1997).

[108]  N. Kobayashi, M. Brandbyge and M. Tsukada, Jpn. J. Appl. Phys. **38**, 336 (1999).

[109]  E. N. Bogachek, A. N. Zagoskin and I. O. Kulik, Sov. J. Low Temp. Phys. **16**, 1404 (1990).

[110]  P. M. Levy and S. Zhang, Phys. Rev. Lett. **79**, 5110 (1997).

[111]  J. F. Gregg, W. Allen, K. Ounadjela, M. Viret, M. Hehn, S. M. Thompson and J. M. D. Coey, Phys. Rev. Lett. **77**, 1580 (1996).

[112]  D. A. Wharam, T. J. Thornton, R. Newbury, M. Pepper, H. Ahmed, J. E. F. Frost, D. G. Hasko, D. C. Peacock, D. A. Ritchie and G. A. C. Jones, J. Phys. C **21**, L209 (1988).

[113]  B. J. van Wees, H. van Houten, C. W. J. Beenakker, J. G. Willliamson, L. P. Kouwenhoven, D. van der Marel and C. T. Foxon, Phys. Rev. Lett. **60**, 848 (1988).

[114]  J. Callaway, *Quantum Theory of the Solid State*, 2nd ed. (Academic Press, San Diego, 1991).

[115] J. B. A. N. van Hoof, K. M. Schep, A. Brataas, G. E. W. Bauer and P. J. Kelly, Phys. Rev. B **59**, 138 (1999).

[116] H. Imamura, N. Kobayashi, S. Takahashi and S. Maekawa, Mat. Sci. Eng. B **84**, 107 (2001).

[117] M. Brandbyge, Ph.D. thesis, Technical University of Denmark (1997).

[118] R. Schilling, Phys. Rev. B **15**, 2700 (1977).

[119] H. P. Bader and R. Schilling, Phys. Rev. B **19**, 3556 (1979).

[120] L. Berger, J. Apll. Phys. **55**, 1954 (1984).

[121] G. G. Cabrera and L. M. Falicov, Phys. Stat. Sol. (b) **61**, 539 (1974).

[122] R. J. Soulen Jr., J. M. Byers, M. S. Osofsky, B. Nadgorny, T. Ambrose, S. F. Cheng, P. R. Broussard, C. T. Tanaka, J. Nowak, J. S. Moodera, A. Barry and J. M. D. Coey, Science **282**, 85 (1998).

[123] S. K. Upadhyay, A. Palanisami, R. N. Louie and R. A. Buhrman, Phys. Rev. Lett. **81**, 3247 (1998).

[124] A. F. Andreev, Sov. Phys. JETP **19**, 1228 (1964).

[125] M. J. M. de Jong and C. W. J. Beenakker, Phys. Rev. Lett. **74**, 1657 (1995).

# Chapter 5

# Applications of Magnetic Nanostructures

*Stuart S. P. Parkin*

## 5.1 Introduction

Magnetic materials find wide application for an enormous variety of uses ranging from electrical motors and generators and magnetically levitated trains, to acoustic speakers, compasses, recycling centers and metal detectors [1]. However, perhaps one of the most important applications of magnetic materials today is their various uses for storing digital information as magnetic bits on magnetic tapes and disks. The ability to store and access vast amounts of digital information is revolutionizing the fundamental fabric of knowledge itself. Already a significant amount of information is in digitized form mostly off-line in compact disks and magnetic tape. Digital data, which is accessible in file and storage systems connected to local or global electronic networks, is rapidly increasing. A comparatively small portion of this information is in perhaps the most useful form of digital data, namely on-line searchable and quereable databases but such databases are growing at exponential rates.

In this chapter we will review, in particular, the use of magnetoresistive magnetic multilayered structures in magnetic recording disk drives. This discussion will necessarily be very limited in scope. However, there are a number of excellent recent books and review articles describing magnetic storage technologies in great detail [2-8] Another application for magnetic materials, which has long been studied, is their potential use for storing information in magnetic memory cells in solid-state memories. Whilst a number of magnetic memory technologies have been explored ever since the 1950s none of these are in much use today. Ferrite core memory was an important technology in the 1950s and 1960s but was not scaleable and was surpassed by alternate technologies. Surprisingly, although ferrite core memory is a very

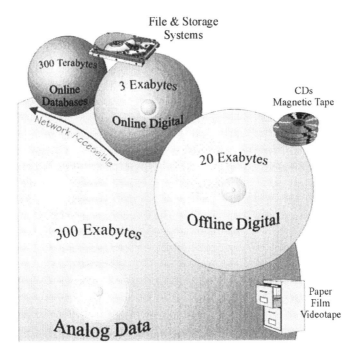

Figure 5.1: Diagram showing estimate of amount of information stored in analog and digital format in 2000. (Note that 1 exabyte is $10^{18}$ bytes.) Most information is still stored in analog form (300 exabytes) but increasing amounts of data are found in digital format either off-line or in information storage devices attached to local or global networks [source: IBM Research].

low performance and extremely expensive memory, it is still in use today for certain highly specialized military applications (satellites and missiles) where a non-volatile radiation hard memory is needed. Magnetic bubble memory was extensively explored until comparatively recently but was not ultimately successful. These memory devices are briefly described in this chapter but more information can be found elsewhere [9-12] Developments in magnetoresistive magnetic multilayers over the past ten years promise to perhaps fulfill the dream of a high performance magnetic random access memory (MRAM) in the coming years [13]. The concept underlying this novel high-performance MRAM will be described towards the end of this chapter.

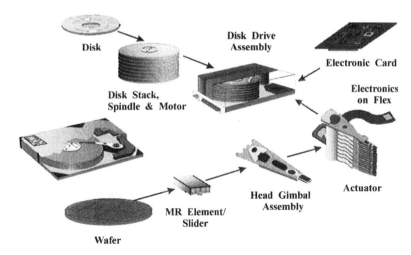

Figure 5.2: Schematic drawing of typical magnetic recording disk drive and its major components. The disk drive contains a series of platters or disks which can be rotated by means of a motor.

## 5.2   Magnetic recording

The basic structure of a typical magnetic recording hard disk storage drive is illustrated in Fig. 5.2. The device consists of one or more rigid disks or platters which are stacked together about a common spindle which is driven by a motor capable of rotating the disks at great speed (up to 15 000 rpm today). Data is stored as magnetized regions or bits within a thin magnetic film within a complex multi-layered thin film structure comprising the magnetic storage medium. The medium is deposited on one or both sides of the disk surface. The outer diameter of the disk is limited by the form factor of the hard disk drive: today most disks have a diameter of either approximately 2.5 or 3.5 inches. Data is written and read from the disk surface by a recording head which is suspended just above the surface of the disk and can be moved across the disk surface approximately radially by the head actuator. As shown in Fig. 5.2, the recording head is comprised of a slider which is formed from a very hard ceramic material, typically comprised of aluminum, titanium and carbon. The reading and writing elements of the head are formed by thin film deposition and lithographic patterning techniques [14, 15]. In a first step the thin film read/write elements are formed on the ceramic wafer, which is then sawed into individual sliders. A very important process then takes place in which the surface of the slider is mechanically ground and polished to provide the air-bearing surface (ABS) on which the slider flies above the disk surface. The ABS is patterned to control the air flow between the underside of the

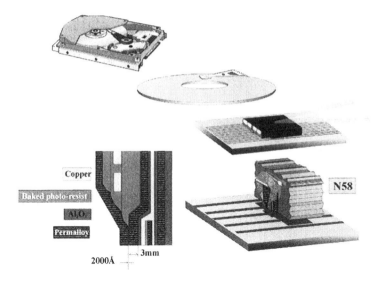

Figure 5.3: Schematic drawing of a magnetic recording disk drive showing one of the disks within the drive and three views of the recording head by which means of which information is written and read to the magnetic storage medium on the disk. The black element is a ceramic slider which flies close above the disk surface. The read and write head, can be seen at the rear face of the slider (the squares indicated as (▢) are the contact pads). Also shown is a more detailed view of the head in which the write head coils (▢) can be seen together with the patterned top pole of the write head ▩. A highly magnified cross-section is shown at the left of the figure.

slider and the disk, when the disk is rotating, so that the closest distance between the ABS and the disk surface is maintained at a constant separation or fly-height, typically today in the range of just 100-200 Å. The air bearing is designed so that the rear edge of the slider is tilted up such that the recording head built on this rear face is brought close to the disk surface.

As is shown schematically in Fig. 5.3, the write head is composed of a series of copper coils (shown as ▢) wrapped between two magnetic poles (shown as ▩) essentially creating a miniature electro-magnet. The poles, which are quite thick ($\sim$ micron), are formed by electro-deposition of a soft magnetic material, typically a nickel-iron alloy. The separation of these poles, the write gap, and the saturation magnetization of the poles, will determine the magnitude and distribution of the magnetic field provided by the write head at the disk surface and thus the size of the written magnetic bits. Obviously the detailed design of the write head is very intricate. One of the most important design considerations is in controlling the magnetic state of the write head poles, not simply when the head is activated, but when the head is not in use so that the write head does not disturb the written bits it will be flying over. Understanding the detailed micromagnetics and control of the magnetic domain structures within the write head magnetic poles is critical [16, 17]. The maximum field which can be provided by the write head is ultimately limited by the saturation magnetization of the pole material. Thus the maximum possible field strengths, which could theoretically be provided by an Co-Fe alloy, are about 2.3 Tesla. Another important consideration is the efficiency of the write head so that the current through the write head coils needed to saturate the poles is the smallest possible. The write head poles must also have high permeability throughout the frequency range of interest, high electrical resistivity to reduce eddy currents, high corrosion and wear resistance, and low magnetostriction (to prevent changes in properties due to stress induced in the manufacturing process) [2].

The write head described above is a longitudinal recording head whereby the magnetic moment of the disk storage medium is in-plane and the write head provides an in-plane fringing magnetic field concentrated near the write gap. The write head fields fall off very quickly with distance from the head gap region. In particular, the write field will vary throughout the thickness of the magnetic storage medium, thereby broadening the transition regions. This is one of the reasons why the thickness of the storage medium and the head-disk spacing must be scaled to ever-smaller dimensions as the bit size is reduced. All magnetic recording drives today are based on longitudinal recording although there is also considerable interest in perpendicular recording which may become important in the future, as discussed in Sec. 5.4.1.

The magnetic bits are written in tracks around the circumference of the disk (see Fig. 5.3). The bits are approximately rectangular in shape as shown in Fig. 5.4 with a bit aspect ratio (the track width divided by the bit length) of between 10 and 20. It is likely that the bit aspect ratio will be reduced in the future. The read head actually detects magnetic flux from the transition regions between oppositely recorded bits. Until 1991 the flux was detected by simply measuring the voltage in the write head coils inductively generated as the head was passed over the transition regions. Since the magnitude of the signal depends on the velocity of the head, the signal strength increases from the inner to the outer tracks, which is a significant disadvantage. There are many other advantages to having separate read and write heads since these can be separately engineered for optimum reading and writing processes. Moreover, the signal provided by an inductive read head is inferior to that provided by read heads based on magnetoresistance (MR). In 1991, read heads based on the anisotropic magnetoresistance (AMR) effect was introduced by IBM [18]. This will be discussed in more detail in a later section but an MR sensor is shown schematically in Fig. 5.4.

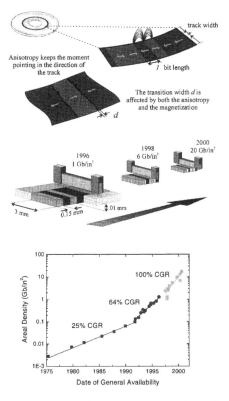

Figure 5.4: Schematic drawing of magnetic recording disk showing a track of recorded magnetic bits and the bit length and track width of each bit. The net magnetic moment of the recorded regions or bits are parallel to the circumference of the disk. The size of the magnetic bits determines how much data can be stored per unit area of the media. The areal density has been increasing at a tremendous pace in recent years as shown schematically in the middle of the figure. At the bottom of the figure is a plot of areal density as a function of general availability of these products in the marketplace.

The cross-section view in Fig. 5.3 shows a cartoon of a thin film MR sensor sandwiched between two magnetic shields (shown as ▨) the upper of which also forms the lower pole of the write head. Note that the spatial sensitivity of the head for both reading and writing is intimately related to the write and read gaps (the latter is the separation of the shields at the ABS). It is also useful to be able to design read and write heads so that the read sensor is only sensitive to the middle portion of the write track to avoid track edge irregularities in the written bits and to relax the tolerance on the track following servoing of the read head.

Magnetic recording technology has evolved at a tremendous pace in recent years. From the mid 1950s when IBM introduced the first hard disk drive, the areal density, the number of magnetic bits per unit area has increased from just $2\,\mathrm{kbits/in^2}$ to more than $20\,\mathrm{Gbits/in^2}$ in 2000. IBM's original hard disk drive, the RAMAC, had fifty 24 inch diameter disks but had a capacity of only 5 Mbytes. Today's disk drives can store 10-15 Gbytes on a single disk only 3.5 inch in diameter! Moreover the data rate has vastly increased from $70\,\mathrm{kbits/s}$ in 1956 to several hundred Mbits/s in 2000. From 1956 to 1991 the areal density grew at a compound growth rate of about 25%/year and from 1991 to about 1997 the areal density grew at a compound growth rate of just over 60%/year. The increased rate in 1991 was caused by several factors, including the introduction of the AMR head, an increased competition in the market place, and a transition to smaller disks which enabled a more rapid evolution of the technology [19]. Since IBM first introduced MR read sensors based on spin-valve giant magnetoresistance (GMR) structures in 1997 the rate of increase of areal density has exploded to between 100 and 200% per year. Just as the areal density has increased the price per bit of storage has dramatically decreased from several thousand dollars per Mbyte in 1980 to about 1 cent per Mbyte today. Indeed one of the most important reasons why magnetic disk drive storage continues to be such an important technology is its very low cost, particularly when compared to the much higher cost of solid-state memory. The technology does have some inherent drawbacks compared to solid-state memory. A particular disadvantage is the long access time of several msec to access the first magnetic bit of information because of the need to rotate the disk, a fairly massive object, to the head. The data rates are also lower than possible with solid-state memory. Disk drives also consume significant power, are noisy and are less reliable than non-mechanical storage.

The evolution of magnetic recording has largely been a matter of scaling all dimensions of the head and media proportionately [19]. However, the areal density cannot continue to increase at the current rate. Magnetic recording will reach limits imposed by the super-paramagnetic effect, the head-disk spacing and switching speed limitations in the head and media.

## 5.2.1 Magnetic media

The gap between the magnetic read/write head and the spinning disk surface is known as the fly height and is about $150\,\text{Å}$ for an areal density of $\sim 30\,\mathrm{Gbit/in^2}$. This close proximity between the head and the disk demands that the disk surface be extremely smooth with a root mean square (rms) roughness of less than a few nm. Traditionally, disk substrates were made from

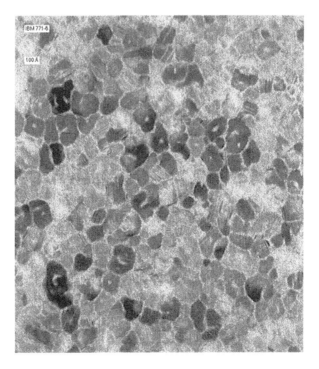

Figure 5.5: Transmission Electron Micrograph of a Co-Cr-Pt-B magnetic media suitable for supporting an areal density of more than 35 Gb/in$^2$ [courtesy of P. M. Rice, M. Doerner and K. Rubin (IBM Almaden Research Center and Storage Technology Division)].

an aluminum-magnesium alloy (Al-4 wt% Mg). A hard layer of an amorphous nickel-phosphorous alloy (Ni-P), which contains paramagnetic Ni, is plated, by an electroless process, on the Al-Mg substrates and is polished to achieve the necessary smoothness. Recently, glass substrates made by a floatation process which yields extremely smooth surfaces (rms roughness of ~5 Å), are increasingly used especially for high performance applications. The substrate, whether Al-Mg or glass, is coated with a sputter deposited underlayer or series of underlayers to control the crystallographic orientation and structural morphology of the magnetic layer which is deposited on top of the underlayer. A commonly used underlayer is Cr or a Cr-V alloy layer perhaps a few hundred angstroms thick. The magnetic layer is typically made from a Co based alloy which may contain as many as three or four other elemental constituents which are used to control the magnetic and physical properties of this layer including the coercivity and squareness of its magnetic hysteresis loop and its corrosion

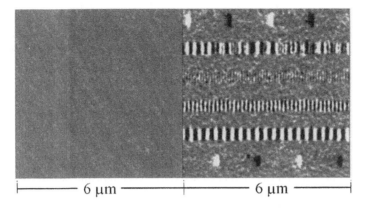

|———— 6 μm ————|———— 6 μm ————|

Figure 5.6: The left panel shows an atomic force microscopy image of the surface morphology of a typical magnetic disk. The disk is very smooth with an rms roughness of ∼ 8-12 Å. The right panel shows a magnetic force microscopy (MFM) image of a bit pattern written at various kfc/mm. The top and bottom rows are written at 0.72 kfc/mm. The second row from the top has a Pseudo-Random bit pattern. The third row has a bit pattern of 21.4 kfc/mm which corresponds to an areal density of 35 Gb/in². The fourth row has a bit pattern of 14.3 kfc/mm. The fifth row has a bit pattern of 7.2 kfc/mm.

resistance [20]. The crystallographic orientation of the magnetic layer plays a major role in determining its magnetic properties. Co alloys usually have an hexagonal close packed structure and consequently a large magnetocrystalline anisotropy with the easy magnetic anisotropy axis along the c-axis. Thus for longitudinal recording it is necessary to orient the c axis in or close to the plane of the film. This is accomplished by choice and growth of the underlayers and by taking advantage of preferred crystallographic relationships of the Co alloy structure and that of the underlayer [20]. Whilst a Cr underlayer is often used with NiP/AlMg substrates, other seed layers such as NiAl are used for glass substrates. The Co alloy magnetic layer is typically ∼100 to 300 Å thick with a Co concentration typically in excess of 70%. About 10-15 atomic percent Cr is used to increase coercivity, to provide improved corrosion resistance and to magnetically decouple individual grains. The Cr is known to partially segregate to the grain boundaries so by proper choice of growth and the Cr content a non-magnetic Co-Cr alloy is formed at the grain boundaries. The Cr also forms a thin surface oxide layer, which acts as a barrier to further oxidation of the magnetic material. The Pt concentration in the alloy is similar to the Cr concentration and is included to increase the coercivity of the magnetic layer. In some cases additives such as boron are also added to the magnetic layer to improve Cr segregation and to further increase coercivity [21].

The microstructure of a typical magnetic media is shown in the through-foil transmission electron microscopy (TEM) image in Fig. 5.4. This Co-Cr-Pt-B quaternary alloy film is capable of supporting an areal density of more than $35\,Gb/in^2$. The image shows clearly that the media is polycrystalline and that there is a narrow size distribution of many fine grains each approximately $85\,Å$ in size. The interior of each grain is a Co rich hexagonal close packed phase whereas the inter-granular phase is an amorphous Cr-B rich phase. This Cr-B phase is non-magnetic and thus magnetically isolates each grain from its neighbors.

Figure 5.6 illustrates some typical bits written on the magnetic media shown in Fig. 5.5. The left panel shows the surface morphology of the magnetic media as measured by an atomic force microscope (AFM) and demonstrates the very smooth surface of this disk. The right panel in Fig. 5.6 shows a magnetic force microscopy (MFM) image of bit patterns written at several different flux changes per mm (fc/mm) and correspondingly different areal densities. The third row shows a bit pattern corresponding to an areal density of $35\,Gb/in^2$ which is nearly two times higher than the highest areal density in magnetic disk drives today. Note that the magnetization within each bit is approximately parallel to the width of the bit along the circumference of the disk. Since the MFM senses field gradients the MFM only detects boundaries (transitions) between bits written in opposite directions. Note that the bits are approximately rectangular in shape. Historically the bit aspect ratio, the ratio of the bit length to its width, has been about 16:1 but is likely to decrease in the future as the areal density further increases.

The magnetic layer has to be protected from atmospheric oxidation and from the read/write head. An overcoat of carbon with a film thickness of $\sim150\,Å$ is used. This carbon film is also sputter deposited on to the disk surface. The hardness of the carbon overcoat can be adjusted by varying the hydrogen content of the film. Both the disk surface and the recording head surface are very smooth and hence the van der Waals interaction particularly during landing and parking of the read/write head may cause the head to stick to the disk surface. This phenomenon is called stiction and must be prevented. This is primarily achieved by coating the disk surface with a thin lubricant layer by a dipping process. A typical lubricant is a fluoro-polymer with a well defined molecular weight and with a thickness on the disk surface of $\sim20\,Å$. The lubricant molecules may have functional end groups if bonding to the carbon overcoat is desired. In older drives the surface of the disk substrate was mechanically circumferentially textured to provide a circumferential magnetic anisotropy along the data tracks. Another benefit of this texturing was a reduced interaction between the head and the media because of the rough disk surface. However, as the fly height shrinks to ever smaller dimensions, the disk surface must correspondingly be made smoother making stiction more likely. This is such an important problem that in modern drives the recording head is parked either on special landing zones or off the surface of the disk as discussed below.

## 5.2.2   Limits to magnetic recording

In some of the grains in Fig. 5.5 lattice fringes can be observed. The orientation of the fringes and the corresponding $c$ axis and magnetic easy axis of these grains varies randomly from grain to grain. Moreover, the magnitude of the switching field of an individual grain will also depend on the volume and shape of the grain. Note that if the magnetization of the grain rotates coherently, and the moment of the grain is decoupled from its neighbors, then the switching field is given by $H_c \sim K_u/M_s$, where $K_u$ and $M_s$ are the uniaxial anisotropy and saturation magnetization, respectively. Thus magnetic recording relies on statistically averaging over very many grains to avoid large variations in signal from bit to bit. In order to achieve reasonable signal to noise ratios (SNR) one bit is typically comprised of several hundred to a thousand grains. Note that the SNR is related to the number of grains $N$ as $\sim 10 \log N$. Thus, as the recording density is increased the size of the grains within each bit must be decreased. This is achieved by fine-tuning the various layers within the magnetic media. However, this process cannot continue indefinitely because eventually the grains will become super-paramagnetic, whereby the magnetic moment of the grain is likely to undergo spontaneous switching of its magnetic state by assistance from thermal energies. The thermally assisted reversal is due to random magnetic field fluctuations arising from an interplay of magnons and phonons [22]. The higher is the magnetic anisotropy of the grain $K_u V$ the greater is the stability of the grain's magnetic moment against thermal fluctuations ($V$ is the volume of the grain). Magnetic disk drives are designed such that magnetic bits be capable of storing information for at least 10 years. This requires that $K_u V/k_B T > 40\text{-}50$ where $k_B$ is Boltzmann's constant and $T$ is the absolute temperature. The ultimate areal density achievable may ultimately be limited by the super-paramagnetic effect. Indeed, there have been numerous theoretical predictions [23] of the demise of magnetic recording at various areal densities due to super-paramagnetism but none have so far come to pass!

An obvious means of surmounting the super-paramagnetic effect is to increase the magnetic anisotropy of the magnetic grains. Since typical current magnetic media have grains with a log-normal size distribution, engineering media with narrower grain size distributions is also helpful [24]. A caveat, however, is the increased coercivity ($H_c$) of the media, which as mentioned above, is directly related to $K_u$ ($H_c = K_u/M_s$). Although high anisotropy media films can show coercivity exceeding 10 000 Oe this far exceeds the write field capability of current recording heads. Many novel approaches are being explored to circumvent this issue. These approaches include perpendicular

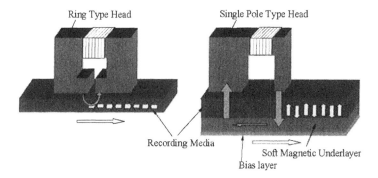

Figure 5.7: Comparison between longitudinal and perpendicular recording. In longitudinal recording the bit size is determined by the gap between the two poles of the write head. In perpendicular recording the bit size is determined by the size of pole on the right.

recording, thermally assisted recording, lithographically patterned media, or media consisting of a self assembled, ordered nanomagnet array.

### 5.2.2.1  Perpendicular recording

In perpendicular recording the magnetization within the bits is perpendicular to the magnetic media surface [25]. The magnetic storage medium in this case is deposited on top of a high permeability soft magnetic layer. This layer can be considered as part of the write head structure as it serves to flux close the path between the two poles of the write head (see Fig. 5.7). This pole head - underlayer combination gives much higher fields, up to 80% higher than related heads used for longitudinal recording. Thus this combination allows the use of higher anisotropy $(K_u)$ media. Furthermore, this combination confines edge effects during the writing process. The media requirements are more complex than discussed for longitudinal recording primarily due to the need for the combination of a soft magnetic underlayer and the recording media. It is possible to obtain perpendicularly magnetized media by several approaches. One class of media are polycrystalline granular media discussed above but where, for example, the easy magnetic axis of the hcp Co-Pt-X material is engineered to be perpendicular to the magnetic film. A second class of magnetic multilayered media is comprised of ultra thin layers of ferromagnetic material, e.g. Co or Fe, separated by thin layers of Pd or Pt. The magnetic interface between the Co or Fe and the Pd or Pt layers exhibits a perpendicular surface anisotropy which, by making the magnetic layers very thin, can dominate the in-plane volume anisotropy of the magnetic layers [26]. Ouchi and Honda have reviewed

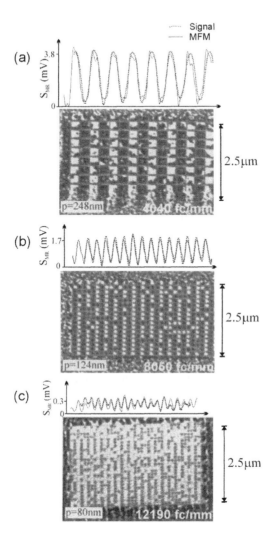

Figure 5.8: MFM images of in-phase written island arrays of various sizes. The linear densities of the written bit-patterns are denoted in the lower right image corners. The particular MFM signals averaged over the column length and the corresponding read-back signals are plotted above the respective image. (a) the island pitch is 248 nm (b) the island pitch is 124 nm (c) the island pitch is 80 nm and the recording density is 19.6 Gb/in$^2$ (from Lohau *et al.* [32]).

recent work on perpendicular media and have shown that granular media of the Co-Cr type exhibit high SNR and low $H_c/4\pi M_s$ whereas multi-layer media exhibits low SNR and high $H_c/4\pi M_s$ [27]. In the granular media there is no coupling between grains. Hence the magnetic reversal process involves the reversal of individual grains without domain wall movement. This yields high SNR but because $H_c/4\pi M_s$ is low the thermal stability is not very good. In the multi-layer media there is cooperative reversal of grains via domain wall movement. Hence the SNR is poor but because of high $H_c/4\pi M_s$ the thermal stability is very good. The transition region between two adjacent bits in perpendicular recording have much lower demagnetizing fields associated with them than for the case of longitudinal recording. The low demagnetizing field does not broaden the transitions and thereby allows the thickness of the media to be much larger than those used in longitudinal recording. Hence, a larger number of grains can be present in each magnetic bit in perpendicular magnetic media. Thus perpendicular recording can potentially support areal densities which are 2 to 5 times greater than longitudinal recording. This mode of recording is however faced with problems associated with excess noise in the soft underlayer. Furthermore, both new write head and read/write electronics are required. It nevertheless appears that perpendicular recording may play an important role as recording densities exceed $\sim$100 Gb/in$^2$.

### 5.2.2.2   Thermally assisted recording

In thermally assisted recording the limitations of conventional write heads to write magnetic bits on very high anisotropy media is overcome by carrying out the writing process at elevated temperatures [28, 29]. A laser beam is used to locally heat the media which temporarily reduces the coercivity of the media in the heated spot. The magnetic bit is then recorded in the heated region with a conventional write head. Since this proposal allows the use of very high anisotropy media there is potential to achieve extremely high areal densities.

### 5.2.2.3   Lithographically patterned media

Patterned media is another scheme to overcome the thermal stability problem as very high areal densities are approached [30, 31]. This involves the creation of single domain magnetic islands, for example, by lithographically patterning a continuous magnetic film. The domains may consist of many grains but these are strongly exchange coupled. Thus there is no need to reduce the grain size to maintain SNR as the bit size is reduced. The coercivity of these single domain magnetic islands is low and well within the limits of the currently achieveable write fields. It is desirable that the easy axis of the islands be aligned and preferably along the perpendicular direction by use of crystalline anisotropy. Figure 5.8 illustrates an example of the use of patterned media.

Figure 5.9: TEM micrograph of a 3D assembly of chemically synthesized FePt nanoparticles with 60 Å diameter. Picture size is $1300 \times 1300$ Å and the particle density is about 20 Tparticles/in$^2$ (from Sun *et al.* [Sun, 2000 #2305]).

In this case, a focused ion beam was used to pattern granular CoCrPt perpendicular anisotropy media into square islands [32]. MFM images of three island shapes with periods of 248 nm, 124 nm, and 80 nm were written with a square bit pattern in phase with the islands. The written bits were detected with a magneto-resistive head and the read-back signal is plotted as a dotted line. This read signal is compared to the MFM signal averaged over the entire column length (solid line). Excellent agreement is found.

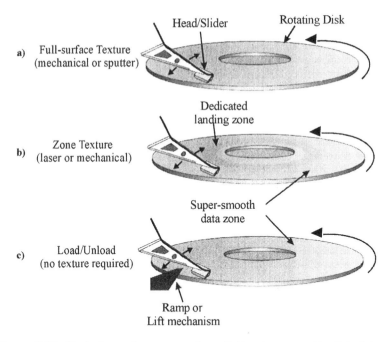

a) Full-surface Texture (mechanical or sputter)

Head/Slider   Rotating Disk

b) Zone Texture (laser or mechanical)

Dedicated landing zone

Super-smooth data zone

c) Load/Unload (no texture required)

Ramp or Lift mechanism

Figure 5.10: Techniques for overcoming stiction in magnetic-disk drives.

### 5.2.2.4  Self assembled nanomagnet array

Superlattices of self-assembled, mono-dispersed FePt nanoparticle arrays have been prepared by Sun *et al.* [33] by simple chemical means. The FePt nanoparticle size could be readily varied in the range of $\sim$30-100 Å with extremely tight tolerances on the sizes of the individual particles. Figure 5.9 shows an assembly of 60 Å particles. Magnetic bits have been recorded in this media with linear densities of $\sim$5000 fc/mm demonstrating the possibility of storing stable magnetic bits at room temperature.

## 5.2.3  Head/disk interface

The magnetic interactions between the read/write head and the disk medium in a magnetic disk drive are very dependent on the spacing between them. In a high performance drive, controlling this spacing precisely is critical to the drive's reliable operation. Modern drives achieve this controlled spacing by using recording heads mounted on the end of small ceramic sliders supported by *air bearings*. The air bearings, which are self-pressurized by the airflow generated by the rapidly spinning disk can be designed to precisely maintain a

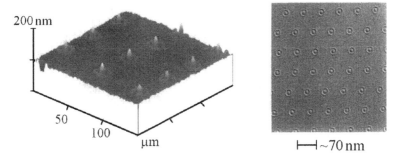

Figure 5.11: Atomic force microscopy of laser textured disk landing zone.

head/disk gap of less than 100 nm at a relative speed of over 30 m per second. To maintain this spacing at these speeds, the slider and disk surfaces must be extremely smooth, and are covered with a thin lubricant layer to improve the durability of the interface. In addition, in most drives, the head rests on the disk surface when the drive is stopped, so the static friction or *stiction* of the head/lube/disk system must be overcome when the drive restarts. The stiction can be powerful enough to damage the head mounting or disk surface. Reducing stiction has been accomplished in various ways (see Fig. 5.10). One possible solution is to texture the entire surface of the disk to a controlled roughness (see Fig. 5.10(a)). Although such a controlled roughness can effectively solve the stiction problem, it limits how far the recording head/medium spacing can be reduced. Other techniques involve a super-smooth data zone, where the head/disk space can be as small as desired, and either a separate dedicated landing zone which has been textured, or a mechanism to load and unload the head from the disk surface as the drive starts and stops, thus keeping the head from ever sitting on the surface while at rest. Of particular interest is one technique for creating a dedicated landing zone in which the disk surface is textured by heating it with a pulsed laser (see Fig. 5.11). This laser texturing produces a region of highly precise, smooth and durable bumps about 40 nm high that reduce the head/disk contact area sufficiently to reduce the stiction to acceptable levels. This laser textured region is used for landing and parking the read/write head. The laser texturing is effective because it reduces the contact area between head and the disk and further reduces incidences of stiction.

The preferred method today to reduce stiction is to remove the head from the surface of the disk when not in use. This method of load/unload is shown in Fig. 5.10(c).

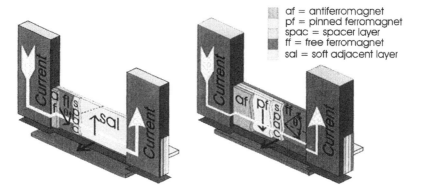

Figure 5.12: Schematic drawing comparing a read head based on anisotropic magnetoresistance (left) with a head based on GMR (right). The sense current is passed through two leads to the multilayered sensing device. The read head is suspended above the magnetic media as shown in Fig. 5.3.

### 5.2.4 Magnetic recording magnetoresistive read heads

Any material whose resistance changes with magnetic field can be used to sense or detect magnetic fields. For magnetic recording read heads the material must be sensitive to magnetic fields in the range of $\sim$10-100 Oe, the magnitude of the fields from the transition regions in the magnetic storage medium. Moreover, the material must have a significant MR at the operating temperature of the MR sensor, which, as discussed below is typically in the range of 100-130 C. It is also useful that the properties of the sensor are not very sensitive to temperature. The MR sensor, which is exposed at the ABS, must also be highly resilient and corrosion resistant. Interestingly, although data is stored in magnetic disk drives digitally the data is read back as an analogue waveform (see, for example [2]). Thus the sensor must give an approximately linear response with magnetic field up to the maximum field sensed. These stringent requirements rule out many materials. The first magnetoresistive read heads were based on the AMR effect but today the vast majority of recording heads are based on GMR spin-valve structures (see Fig. 5.12) as discussed later in this section.

#### 5.2.4.1 Anisotropic Magnetoresistance Sensors

Typical $3d$ ferromagnetic metals or alloys display only small changes in their resistance when subjected to magnetic fields at room temperature [34]. Maximum MR values of about $\sim$5-6% are found in Ni-Co and Ni-Fe alloys although these values can be much larger at low temperatures. In magnetic fields large

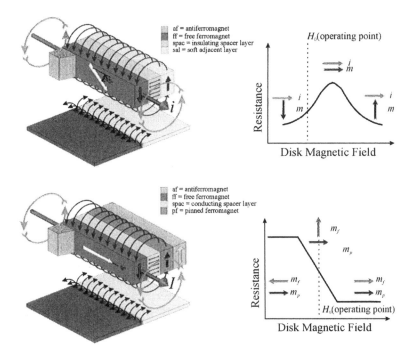

Figure 5.13: Biasing scheme for (top) an anisotropic magnetoresistive reading head and (bottom) a giant magnetoresistive spin-valve head.

enough to saturate the magnetic moment of such metals their resistance primarily depends on the orientation of their magnetic moment with regard to the direction of the sense current passing through the sample. Thus they display an AMR such that their resistance can be written as $\rho = \rho_0 + \Delta\rho\cos^2(\theta)$, where $\theta$ is the angle between the magnetic moment of the sample and the direction of the current [34, 35]. The resistance is typically higher when the magnetic moment of the sample is aligned orthogonal to the sense current. In magnetic fields not large enough to saturate the magnetization of the metal the resistance depends on the detailed magnetic domain structure. In thin ferromagnetic films, the magnitude of the AMR effect becomes even smaller as the thickness of the FM layer is decreased because scattering of the conduction electrons from the outer boundaries of the film increases the resistance of the film. These scattering processes do not give rise to AMR.

In order to obtain a linear response from an AMR sensor the magnetic moment of the sense layer must be arranged to be approximately at an angle of 45° to the sense current in zero magnetic field. A variety of ingenious means have been developed to make this possible [4, 5]. One scheme, shown in Figs. 5.12 and 5.13 (top), is to use a soft adjacent layer (SAL) separated from the AMR sense layer by a thin spacer layer (typically tantalum). The SAL is formed from a very soft ferromagnetic layer which displays no MR and which is typically much more resistive than the sense layer. Thus the basic AMR sensor is comprised of a sandwich of two thin ferromagnetic layers separated by a thin non-magnetic layer. This structure is surprisingly similar to the spin-valve MR sensor discussed below. When current is passed through this sandwich the self-field from the current causes the moment of the SAL to become oriented perpendicular to the ABS (see Fig. 5.13 top). The magnetic poles formed at the top and bottom of the SAL then provide a magnetostastic field which the MR sense layer is subjected to. By simply engineering the relative magnetic moments of the SAL and the sense layer (the SAL moment is larger than that of the sense layer) the moment of the sense layer will become oriented at a particular angle to the sense current. This angle depends not only on the magnetostatic coupling between the SAL and sense layer, bust also on any magnetic coupling between the SAL and sense layer through the spacer layer, for example due to Neel orange-peel coupling [36], as well as the magnetic anisotropy of the sense layer (due to, for example, the intrinsic magnetocrystalline anisotropy of the sense layer material, or from anisotropy in the structure introduced by growing the sense layer in a magnetic field, or by shape anisotropy). Thus by proper engineering of the materials and structure the sense moment can be arranged to be at the required angle with respect to the current during operation and a linear response of the MR sensor is obtained over the required field range (see Fig. 5.13).

### 5.2.4.2   Giant Magnetoresistance Sensors

Scaling of magnetic recording heads with increasing areal density requires reducing the thickness of the sense layer in the read head. For example the thickness of the AMR sense film has to be decreased from approximately $\sim 150$ Å at 1 Gbit/in$^2$ to well below $\sim 100$ Å at densities of $> \sim 5$ Gbit/in$^2$. Since AMR arises from scattering throughout the AMR layer, as the AMR layer thickness is reduced, the AMR effect is reduced and becomes too small to be useful for recording read heads at densities of $\sim 5$ Gbit/in$^2$.

As discussed in earlier chapters certain magnetic multilayers, containing very thin ferromagnetic layers can display very large or *giant* changes in resistance with magnetic field [37-40] The largest GMR effects have been found in multilayers, prepared by sputter deposition, composed of alternating Co and Cu layers, each just four or five atomic layers thick. In polycrystalline Co/Cu multilayers GMR effects of more than 70% have been found at room temperature [41]. In crystalline Co/Cu multilayers even larger effects have been found. Figure 5.14 shows the largest yet reported GMR effect at room temperature of more than 110% (Parkin, unpublished).

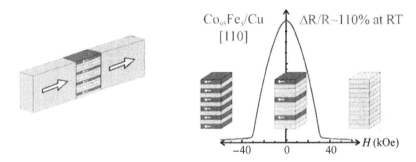

Figure 5.14: Schematic representation of GMR measured with current in the plane of the layers together with a resistance vs field curve for a sputtered epitaxial [110] oriented Co/Cu multilayer for field applied along the [011] direction in the plane of the superlattice.

As discussed in earlier chapters the origin of the giant magnetoresistive effect is quite different from that of AMR. GMR is found in multilayered and other inhomogeneous magnetic structures in which the magnetic layers (or other entities such as magnetic granules in magnetic granular metals [42]) are oriented non-parallel to one another for some range of magnetic field, and, such that, with application of a sufficiently large magnetic field, the magnetic moments of the layers (or entities) become oriented parallel to one another. It is the change in the magnetic configuration which affects the scattering of the conduction electrons propagating between the magnetic layers or entities and which thereby gives rise to GMR. In Co/Cu multilayers, for certain thicknesses of Cu, the moments of the Co layers are arranged antiparallel to one another in small fields because of an antiferromagnetic (AF) coupling of the Co layers mediated via the Cu spacer layers. When a magnetic field is applied, large enough to overcome the AF interlayer coupling, the Co moments becomes aligned parallel to each other and to the applied field.

Polycrystalline Co/Cu multilayers are usually 111 textured for thin Co and Cu layers, although the texture changes to 100 for thick Cu layers [43] or when the multilayer is grown on thick Cu buffer layers [44]. Polycrystalline multilayers usually display little in-plane magnetic anisotropy. Consequently, the resistance of such multilayers typically varies continuously with magnetic field independent of the orientation of the magnetic field in the plane of the sample [41, 45, 46]. For strongly antiferromagnetically coupled multilayers, as the magnetic field is increased, the angle between neighboring magnetic layers, ~180° in small fields, smoothly decreases until at magnetic fields large enough to overcome the antiferromagnetic interlayer exchange coupling the magnetic moments becomes aligned parallel to the magnetic field and to each other. When multilayers are crystalline and have significant magnetic anisotropy the

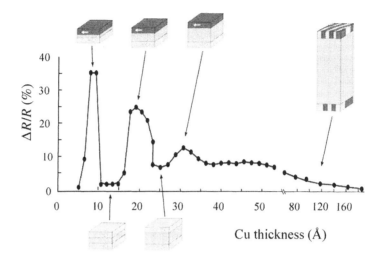

Figure 5.15: Room temperature saturation magnetoresistance vs Cu spacer layer thickness for a series of Co/Cu multilayers [41]. The magnetic state of the multilayers are shown schematically for various Cu spacer layer thicknesses (only two Co layers are shown).

dependence of resistance on magnetic field is more interesting and can display quite unusual behavior [47].

As shown in Fig. 5.15 for Co/Cu multilayers, the magnitude of the GMR effect was discovered to oscillate as a function of increasing spacer layer thickness between the ferromagnetic layers [45, 46]. The oscillation in saturation MR is related to an oscillation in the sign of the interlayer exchange coupling, between antiferromagnetic (AF) coupling and ferromagnetic (F) coupling, as the spacer layer thickness is varied. This is shown schematically in Fig. 5.15. These oscillations in both MR and interlayer coupling were first observed in sputter deposited Fe/Cr and Co/Ru multilayers [45]. The coupling via Cu, Cr, Ru and other transition and noble metals was found to be long-range and presumed to be of the RKKY type. In polycrystalline Co/Cu multilayers the oscillation period is ~10 Å although in crystalline multilayers the period depends on the crystal orientation of the multilayer [48]. The first observation of oscillatory interlayer coupling in transition metal multilayers was in Fe/Cr and Co/Ru sputtered polycrystalline multilayers [37]. Subsequently it was shown that oscillatory interlayer coupling is exhibited by nearly all of the 3*d*, 4*d*, and 5*d* non-ferromagnetic transition and noble metals [49]. Later, oscillatory coupling was observed in single-crystalline Fe/Cr and Co/Cu films grown by evaporation techniques in ultra-high vacuum chambers [48]. For (100) Fe/Cr and (100) Co/Cu the interlayer exchange coupling oscillates with

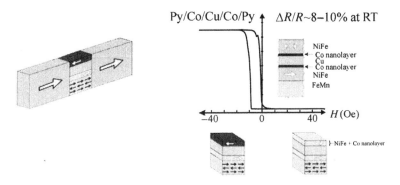

Figure 5.16: Schematic representation of giant magnetoresistive spin valve structure together with typical magnetoresistance vs field curve (circa 1991) for the structure shown in the insert to the data.

Cr and Cu spacer layers with two superposed oscillation periods, one long and one short [50, 51]. The magnitude of the oscillation periods for noble metal spacer layers can be well accounted for by examination of the Fermi surfaces of the noble metals. The oscillation periods are related to wave-vectors which span or nest the Fermi surface [52, 53].

### 5.2.4.3   Giant magnetoresistance spin-valve sandwiches

Although GMR multilayers exhibit very large MR these are not useful structures for MR recording heads because the saturation fields are too large (see Fig. 5.14). However, useful devices can be formed based on GMR and taking advantage of oscillatory interlayer coupling, which are very sensitive to small magnetic fields. This device, the spin-valve (see Fig. 5.16), is composed of two thin ferromagnetic layers separated by a thin Cu layer [54]. The device relies on the exchange-biasing [55, 56] of one of the ferromagnetic layers to magnetically pin this layer. This effect, of ancient origin, is described schematically in Fig. 5.17.

The magnetic hysteresis loop of a ferromagnetic layer is centered symmetrically about zero field. However, certain combinations of thin F and AF layers display hysteresis loops which are displaced from zero field by an *exchange bias* field [57]. The origin of the effect is related to an interfacial exchange interaction between the AF and F layers and the fact that the magnetic lattice of the AF layer is essentially rigid, and little perturbed by even large external magnetic fields. Assuming the simplest possible AF structure of successive ferromagnetically ordered atomic layers whose moments alternate in direction from one layer to the next, one can readily appreciate that the uncompensated magnetic moment in the outermost AF layer at the AF/F interface will give

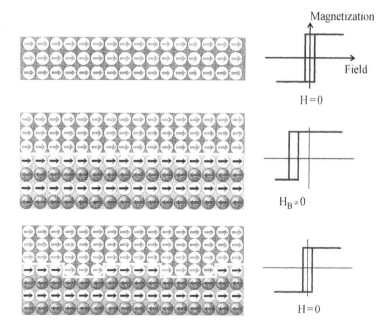

Figure 5.17: Schematic diagram of exchange biasing. Idealized magnetization vs field hysteresis loop for a F thin film (top) and a F thin film exchange biased by an AF layer (middle). The bottom shows the absence of exchange bias for an imperfect F/AF interface.

rise to a exchange field which the F layer is subjected to. A long standing puzzle is why any exchange bias field is observed at all since one supposes that the interface between the F and AF layers is rough on an atomic scale [58]. As shown in Fig. 5.17, if the interface consists of atomic terraces, whose length is less than the exchange length in the F metal, there will be no net exchange anisotropy field. Note that similarly, if the AF layer is composed of randomly oriented magnetic domains, then this alone would quench the exchange bias field. In order to establish an exchange bias field the AF layer is usually deposited on a magnetized F layer such that the interfacial exchange anisotropy leads to a preponderance of domains in the AF layer contributing to a net exchange bias field. Alternatively, by heating the F/AF combination above the so-called blocking temperature of the AF layer where the AF spin system is no longer rigid, and subsequently cooling the bilayer couple in a magnetic field, an exchange bias field can be established in the direction of the applied field. This is a useful method to orient the exchange bias field in different directions in different magnetic layers in more complicated magnetic structures.

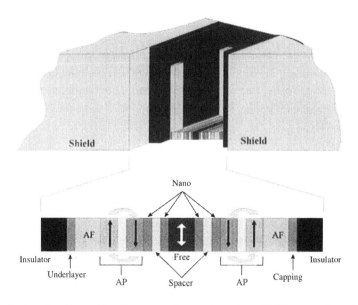

Figure 5.18: A schematic diagram of a spin-valve GMR recording head of the future comprised of two exchange biased pinned F layers on either side of a F free layer. All the F layers have thin interface layers (shown as nano-layers) designed for maximum interfacial spin-dependent scattering. The two pinned F layers are comprised of antiparallel (AP) coupled F bilayers, coupled antiferromagnetically by ultra thin Ru layers [Parkin, 1990 #325]. There may also be some advantages in forming the free layer from an AP structure.

By using AF layers with different blocking temperatures different, F layers can thereby be exchange biased in different directions. This is useful for engineering magnetic structures for various applications. A variety of models have been proposed to account for an exchange bias field even in the presence of rough interfaces [58, 59].

By combining an exchange biased F layer (a pinned F layer) with a simple F layer (a free F layer, which forms the sensing layer) it is thereby possible to engineer the magnetic moments of the two F layers to be either parallel or antiparallel to one another as a function of magnetic field without relying on interlayer exchange coupling (Fig. 5.16). Well-defined magnetic states of the sandwich are obtained in small positive and negative fields with the magnetic moments of the pinned and free layers parallel or antiparallel to one another. This leads, via the GMR effect, to a step-wise change in the resistance of the sandwich in small magnetic fields. The magnitude of the GMR effect in such sandwiches, useful for recording head applications, is comparatively small compared to magnetic multilayers and ranges up to about 16-18%. The maximum MR observed in a spin-valve structure with one pinned F layer is just over 20% [60]. The magnitude of the GMR in the sandwich is reduced for various reasons, including that there are only two magnetic layers [39], and that the Cu spacer layer and the magnetic layers themselves are relatively thick leading to increased dilution of the GMR effect [43]. By using additional magnetic layers such that the free FM layer has two exchange biased F layers on either side of it, separated by two Cu spacer layers, GMR values of more than 24% have been obtained at room temperature [40, 61].

The spin-valve structure shown in Fig. 5.16 has been engineered to be particularly useful for MR heads i.e., to give the largest possible GMR at suitably low fields. Ever since the first observation of GMR there has been much debate about the relative contributions to GMR of interface and bulk spin-dependent scattering. Early models emphasized the role of spin-dependent scattering within the interior of the F layers [62] but subsequent work revealed that interfacial scattering is the dominant contribution to GMR [63, 64]. In particular, it was shown that dusting interfaces in a permalloy ($Ni_{81}Fe_{19}$)-Cu-permalloy spin-valve or in permalloy/Cu multilayers with ultra-thin Co layers dramatically increased the GMR exhibited by these structures by more than a factor of two [63, 64]. Inserting thin Co layers increases the GMR exponentially with a length scale of just 2-3 Å. Subsequently, it was shown, using inverse photoemission, that the electronic structure of the Co layer evolves over the same length scale [65]. Today nearly all spin-valve GMR recording heads take advantage of interface scattering: the free layer is usually formed from a bilayer comprised mostly of a low magnetostrictive ferromagnetic metal (to reduce the chance of stress-related changes in the magnetic properties of this layer during processing) and a thin interface layer designed for maximum interface scattering and GMR. A second common element in spin-valve structures is the use of a pinned F layer formed from two F layers coupled antiferromagnetically to one another through an ultra-thin Ru layer. This structure, which was invented by Parkin [66, 67], stabilizes the moment of the pinned layer and reduces the magnetostatic coupling between the pinned and free F layers. The reason is that the magnetic moment of the pinned layer can be set arbitrarily from zero up to the total moment of the two pinned F sub-layers. Since the exchange bias is provided by an interfacial coupling energy between the AF and pinned F layers, reducing the net moment of the pinned F layer increases the field required to change the magnetic orientation of the pinned F layer. This makes possible the use of AF layers, such as NiO, which would otherwise not provide sufficient exchange bias field. More importantly the stability of pinned F layers is improved even for AF layers which exhibit strong exchange bias fields. Similarly by using an antiferromagnetically coupled pinned layer (AP layer) the magnetostatic interaction between the pinned and free layer moments can be reduced.

As shown in Fig. 5.13 (bottom) the biasing of a GMR spin-valve recording head is quite different from that of an AMR head. In a GMR head, maximum sensitivity to the magnetic field is obtained when the moment of the sense layer is orthogonal to that of the pinned F layer (see Fig. 5.14). Since the magnetic fields from the transition regions in the magnetic storage medium are largely perpendicular to the medium the moment of the free layer is designed to be oriented approximately parallel to the ABS. This means the moment of the pinned layer must be perpendicular to the ABS. This is arranged by setting the exchange bias field perpendicular to the ABS. The moment of the free F layer is arranged to be parallel to the ABS in the presence of current through the device by carefully balancing the magnetostatic coupling field and the orange-peel coupling field between the pinned and free F layers.

By taking advantage interface scattering and AP layers a GMR spin-valve structure of the future may be much more complicated than the structure shown in Fig. 5.16. A cartoon of a future recording read head is shown in Fig. 5.18.

# 5.3   Non-volatile magnetic memory

## 5.3.1   Introduction

A memory using a magnetic storage element possesses the attractive property of non-volatility, namely the state of the memory is maintained even when power is removed from the memory. A variety of MRAM technologies have been explored over a period of many decades. Originally these involved macroscopic ferrite cores arranged in 2 or 3D arrays [11] but perhaps for the past 20 years or so interest has centered on magnetic thin film MRAM. Early interest centered on magnetic bubble technology involving the storage of information in continuous magnetic films [68] but this was not successful.

## 5.3.2   Ferrite core memory

The first implementation using magnet materials in random access memory can be traced back to the first experiments with Mo/Py wrapped ceramics. In products, the pressed and sintered ferrite core memory (FCM) of the mid 1950s became the standard and was used successfully into the 1970s. A typical product contained a 2D array of magnetic tori, with four thin wires threaded through each torroidal core (see Fig. 5.19). Each torus stores one bit of information. A bit was written by simultaneously applying half the necessary switching current through each of two wires. Where the two write wires intersected, the current would be high enough to inductively flip the magnetization of the ferrite core. Ideally, other cores' magnetizations along these write lines

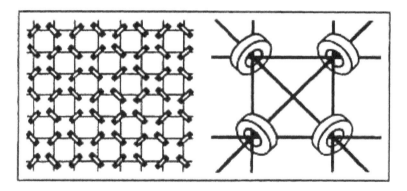

Figure 5.19: 2D array of Ferrite Cores with Write and Sense Wires.

would be left unchanged. A third wire was used to sense the bit. To read a single bit, that bit would be selected and written to a known state. If the core's magnetization flipped, the sense line would sense the field change generated as an induced current. Consequently, the memory was non-volatile, but reading was destructive, and the memory would have to be rewritten after the reading process was complete. To help cancel the interference produced by other bits subject to half the write current, a fourth wire (not shown in Fig. 5.19) was threaded through all of the bits.

The FCM circuits were difficult and expensive to produce. However, ferrite core memory was successful for many years in large part because the technology continued to improve and remained superior to competing technologies. The cores grew smaller, and faster. The core diameter shrank which allowed for higher speed, and less writing current; however, the ferrite core eventually reached its physical limits. The circumferential field of a wire is given by, $H = \frac{.406I}{D}$. Where $D$ is the distance from the wire (cm) and $H$ and $I$ are in oersteds and amperes, respectively. In order to maintain sufficient switching fields, the ratio of the inner and outer diameter needed to remain approximately constant. Manufacturing limits of ferrite prevented ferrite cores getting much smaller than 0.3 mm, a very large size by today's standards.

Another problem arises from the fact that the individual write lines must switch only the core where the write lines intersect, and not be above the disturb threshold of an individual core. If one requires fast, semi-coherent rotation of domains then the half field is near the disturb threshold and consequently will erase unread bits over time. This relegates even the smaller cores, (∼0.3 mm) to less write current and consequently slower switching time on the order of 100 nS [69]. By the late 1960s another faster more scalable technology had entered the storage fray-transistor circuits.

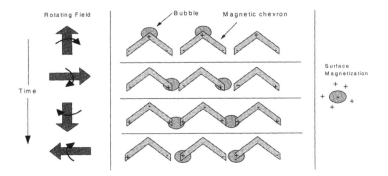

Figure 5.20: Clockwise rotation of magnetic field moving bubbles on permalloy Chevrons [Bobeck and Torre, 1975 #2302].

## 5.3.3 Magnetic Bubble memory

Researchers in the 1970s found a way to store non-volatile bits in a domain of a magnetic medium. Magnetic Bubble memory uses cylindrical magnetic domains, called bubbles to store bits in a shift register. The domains look like bubbles when photographed from above, and this original, misleading name stuck. A magnetic bubble memory module has five major components, a magnetic substrate, a permanent magnetic bias field, an alternating field, patterned magnetic features, and bubble readers.

The magnetic substrate is a canted antiferromagnet (e.g., orthoferrite) with a very small saturation magnetization [70], typically $8\,\text{emu/cm}^3$. In devices, permanent magnets are placed above and below the substrate to provide a bias field to stabilize the perpendicular magnetic domains. The bias field needed to stabilize the domains is on the order of $20\,\text{Oe}$. If the bias field is too large than the bubbles form stripe domains, and if it is too small they collapse.

The bubbles can be created, destroyed and moved by their interaction with the non-uniform field lines from features patterned onto the surface of the magnetic substrate. Uniform fields have no effect on the domains because they have closed domain walls. Experiments with bubble memory used features such as channels or mesa in the substrate, metallic conductors, or most commonly soft magnets deposited on the substrate. The field lines from patterned soft magnet regions in the rotating external field (provided by coils) can be used to push a bubble. Figure 5.20 shows the Intel Magnetics 7110 magnetic bubble memory module using two perpendicular coils excited by a 90° phase shifted triangle waveform to create a rotating magnetic field. Also shown are the bias magnets, substrate and external packaging.

Figure 5.21 shows how bubble motion can be controlled by a rotating external applied magnetic field and chevron-patterned soft magnets deposited on

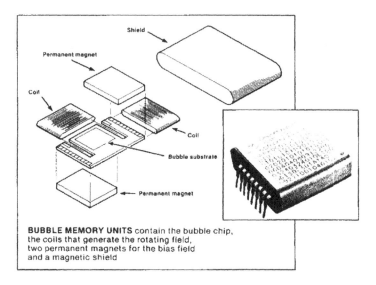

**BUBBLE MEMORY UNITS** contain the bubble chip, the coils that generate the rotating field, two permanent magnets for the bias field and a magnetic shield

Figure 5.21: The Intel Magnetics 7110 magnetic bubble memory module [Source: Intel].

the surface of the magnetic substrate. As the magnetization of the chevron changes polarity with the magnetic field, the bubbles are attracted from one Chevron feature to the next [9]. By patterning the magnets on the substrate, a very long circulating shift register can be formed. The stepping rate for the bubbles in such a register can easily be operated at hundreds of kHz without destroying the bubbles.

Memory devices are produced by creating multiple loops of patterned magnets that form shift registers for the bubbles. Bits are stored basically as either the presence or absence of a bubble in a specific location in the loop register. To increase the signal to noise for detection, the bubbles are replicated from the register and expanded as they approach a detector. Usually, a Hall or MR detector was used to sense the movement of the domains. These replicated bubble are then destroyed. This reading of the replicated domains is non-destructive to the shift register, and when the device is not in use the oscillating magnetic field can be turned off providing non-volatile storage. One major benefit of the magnetic bubble technology is that there are no moving parts in the device so that this allows for robust memory storage unlike removable disks. However, there is an inherent slow access time with the shift register, as all bits must move through the register to read the data again, and there is a limit to the stepping speed. For example, Intel's 7110 has a data transfer rate of 100 Kbit/sec with a maximum storage of 4096 bits in each of

80 loops providing a total 327 680 bit capacity. This was considered enough in the early 1980s!

Magnetic bubble memory was a short-lived technology, which found a few applications in the early 1980s as storage for BIOS or OS in early personal computers. This technology was quickly replaced by the affordable CMOS memory with a battery back-up combined with faster, cheaper hard disk drives.

## 5.3.4 MTJ and GMR Magnetic Random Access Memory

Notwithstanding the ultimate failures of FCM and bubble memories there has continued to be interest in non-volatile random access memories based on magnetic storage elements. In recent years MRAM technologies have favored arrays of individually patterned magnetic storage cells or bits where one bit comprises a magnetic thin film multilayered structure. The magnetic bit is designed to have two stable magnetic states in zero and small magnetic fields which, usually, exhibit two different resistance values representing "0" and "1". Until recently such bits involved the use of the comparatively small AMR effect of conventional ferromagnetic materials arranged in thin film structures. While some of these structures are very ingenious [71], these memories have been, not only of comparatively poor performance, but very expensive, and thus limited in their application.

Replacing AMR bit structures with GMR spin-valve bit structures [72] has some obvious advantages. First, the magnetic states of the GMR bit cells are much simpler (see, for example, Fig. 5.16). Second, the larger GMR effects give rise to larger signals because of the higher MR values. Since, in a first approximation, the time required to read the state of the bit cell depends on how large is the difference in signal between the two states of the cell, this is clearly an advantage. Moreover, the signal from a GMR cell in an appropriately designed MRAM can be sufficiently large that the bit may be read non-destructively without changing its magnetic state. This is not only faster still but consumes less power because the bit does not have to be subsequently re-written. This is clearly an important difference from ferrite core and magnetic bubble memories.

Nevertheless even the larger signals available from GMR structures do not make GMR MRAM attractive for mainstream RAM applications. The reason for this is illustrated in Fig. 5.22. In order to achieve reasonable memory array densities many GMR cells (of number $N$) have to be electrically connected in series which means that the actual signal available when reading one particular cell is $\sim \mathrm{MR}/N$. This signal is not sufficient to make GMR MRAM competitive with conventional dynamic random access memory (DRAM) and static random access memory (SRAM) [73].

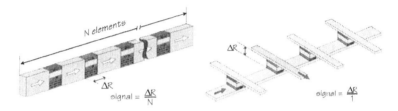

Figure 5.22: Schematic illustration of (a) a number of GMR MRAM cells connected in series along a common bit or line, and (b) a number of MTJ MRAM cells arranged along a common word line (lower line) at the cross-points with corresponding upper bit lines.

By contrast, as illustrated in Fig. 5.22(b), the high MR signal from individual MTJ storage cells can be fully utilized by a novel cross-point MRAM architecture [13, 74], by connecting each MTJ element in series to a switch, for example, a silicon diode. In this case, since current only passes through a single MTJ cell, the available signal when reading that cell is MR/1. With reasonable MR values such an MRAM architecture has the potential to rival that of DRAM in density, and SRAM in speed. The important advantage of an MTJ over a GMR device is thus that current is passed perpendicularly through the MTJ. Thus the electrical contacts to the MTJ cell essentially occupy the same space as the MTJ device itself making the cell very small. The basic structure of the MTJ is otherwise very similar to that described above for the spin-valve GMR recording head. The MTJ cell is comprised of two F layers, an exchange biased pinned F layer and a free F layer, but these layers are separated by an ultra-thin tunneling barrier rather than a metallic spacer layer in the spin-valve structure [13]. Indeed just as for the spin-valve structure the free layer is likely to be comprised of a bilayer where the interface layer is chosen to give maximum TMR and the remainder of the free layer chosen for small magnetorestriction or other properties for optimal magnetic switching characteristics of the free layer or optimal characteristics for processing of the MTJ memory cell. Again, just as for the spin-valve GMR structure there are many advantages to using an AP pinned layer.

In Fig. 5.23, a series of MTJ memory cells are shown at the cross-points between a lower "word" line and an upper "bit" line. These metal lines are conductors through which electrical current can be passed. Typically magnetic memories use a combination of orthogonal bit and word lines [11] in order to be able to individually address each magnetic memory cell to set or "write" its magnetic state. Reading and writing the individual memory cells within the array can be readily be understood by reference to Fig. 5.23. To read the state of a magnetic cell a current is passed along a word line through the

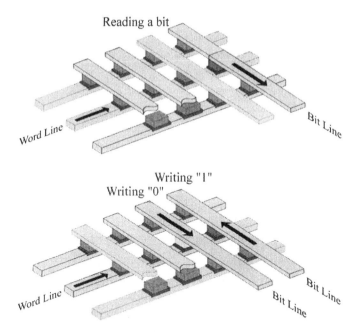

Figure 5.23: Cartoon of a cross-point MTJ MRAM architecture. The magnetic tunnel junction storage cells are located each cross-point of the write and bit lines. The upper and lower cartoons show the processes by which magnetic bits can be read and written.

MTJ itself and along the corresponding word line. Since the resistance of the MTJ device can readily be varied by many orders of magnitude the resistance of the cell can be optimized for reduced power and/or for maximum reading speed. Although, in principle no switch is needed, and a MTJ memory can be constructed with all the MTJ cells in the array connected in parallel, this leads to "sneak" currents passing through all the MTJ cells in the array which consumes more power and reduces the available signal [73, 75].

The state of a selected MTJ cell is written by passing currents simultaneously along the corresponding word and bit lines (only one of the word and bit line currents need be bipolar). The vector combination of the orthogonal magnetic self-fields of the currents or current pulses passed through these lines is arranged such that the magnetic state of the selected memory element at the intersection of the chosen bit and word lines can be appropriately set. On the other hand, the self-fields of these same currents must be such that the magnetic state of the half-selected devices along the same bit and word lines is not altered. Nevertheless these latter cells will be magnetically disturbed

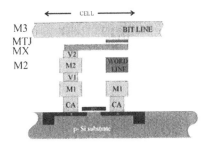

Via/ROM    Cell 1    Cell 2    Cell 3

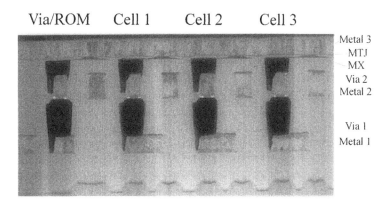

Figure 5.24: The top left figure shows a schematic drawing of a MTJ magnetic memory cell where the MTJ device is connected to an FET fabricated in the silicon substrate using conventional CMOS technology. The MTJ device is connected to the bit line but is not connected to the word line. Conventional CMOS circuits are built beneath the MTJ cells to provide the necessary circuits to read and write individual cells in the cross-point array. A cross-section transmission electron micrograph of 4 MTJ memory cells in a functioning 1 kbit MTJ MRAM array built using 0.25 mm CMOS technology is shown in the bottom of the figure. The word lines can be seen in cross-section in the metal layer 2. These are ∼ 0.4 mm wide. The bit line is shown across the top of the micrograph in metal layer 3. The metal layers 1 and 2 are fabricated using standard Al CMOS techniques. After the CMOS circuits were fabricated within IBM's Microelectronics division the MTJ structure was deposited on top of the CMOS at the IBM Almaden Research Center. The MTJ devices were patterned and the top bit line was fabricated at the IBM T.J. Watson Research Center [Parkin, 1999 #1220][Scheuerlein, 1998 #947].

and it is very important that even after many such disturbances the magnetic state of these cells does not "creep" either to some intermediate state or completely reverse. Unpatterned MTJs without exchange-bias layers appear to be susceptible to creep [76].

Finally, a cross-section transmission electron micrograph of 4 MTJ memory cells in a 1 kbit array where the MTJ cells were integrated with conventional CMOS read and write circuits is shown in Fig. 5.24 [13, 73]. This prototype has been used to demonstrate the feasibility of integrating magnetic tunnel junction magnetic memory storage cells CMOS circuits. The prototype also was used to demonstrate the potentially very high performance of such a magnetic memory. In particular it was shown that an individual MTJ memory cell could be read and written in less than 2.5 nsec which are comparable speeds to that of fast SRAM [73]. The simple cross-point architecture is consistent with very high densities because the size of the cell is small so that densities comparable to DRAM are feasible. Thus we conclude that magnetic tunnel junction random access memory promises a solid-state memory which is not only fast and dense but non-volatile. It will take several years to see whether this promise is realized.

# References

[1] J. D. Livingston, *Driving Force: The natural magic of magnets* (Harvard University Press, Cambridge, 1996).

[2] K. G. Ashar, *Magnetic Disk Drive Technology: Heads, Media, Channel, Interfaces, and Integration* (IEEE Press, New York, 1997).

[3] H. N. Bertram, *Theory of Magnetic Recording* (Cambridge University Press, Cambridge, 1994).

[4] P. Ciureanu, in *Thin Film Resistive Sensors*, edited by P. Ciureanu and S. Middelhoek (Institute of Physics Publishing, Bristol, 1992), p. 253.

[5] J. C. Mallinson, *Magneto-Resistive Heads* (Academic Press, 1995).

[6] C. D. Mee and E. D. Daniel (McGraw-Hill, New York, 1996).

[7] C. D. Mee and E. D. Daniel (McGraw-Hill, New York, 1996).

[8] F. Jorgensen, *The complete handbook of magnetic recording* (TAB Books, New York, 1996).

[9] A. H. Bobeck and E. D. Torre, *Magnetic Bubbles* (North-Holland, Amsterdam, 1975).

[10] S. Middelhoek, P. K. George and P. Dekker, *Physics of Computer Memory Devices* (Academic Press, London, 1976).

[11] R. E. Matick, *Computer storage systems and technology* (John Wiley & Sons, New York, 1977).

[12] D. A. Thompson, L. T. Romankiw and A. F. Mayadas, IEEE Transactions on Magnetics MAG-**11**, 1039 (1975).

[13] S. S. P. Parkin, K. P. Roche, M. G. Samant, P. M. Rice, R. B. Beyers R. E. Scheuerlein, E. J. O'Sullivan, S. L. Brown, J. Bucchigano, D. W. Abraham, Yu Lu, M. Rooks, P. L. Trouilloud, R. A. Wanner, and W. J. Gallagher, J. Appl. Phys. **85**, 5828 (1999).

[14] R. E. Fontana, IEEE Trans. Magn. **31**, 2579 (1995).

[15] R. E. Fontana, S. A. MacDonald, H. A. A. Santini and C. Tsang, IEEE Transactions on Magnetics **35**, 806 (1999).

[16] J. C. Slonczewski, IEEE Trans. Magn. **26**, 1322 (1990).

[17] J. C. Slonczewski, B. Petek and B. E. Argyle, IEEE Trans. Magn. **24**, 2045 (1988).

[18] C. Tsang, M. M. Chin, T. Togi, *et al.*, IEEE Trans. Magn. **26**, 1689 (1990).

[19] D. A. Thompson and J. S. Best, IBM J. Res. Develop. **44**, 311 (2000).

[20] K. E. Johnson, J. Appl. Phys. **87**, 5365 (2000).

[21] N. Tani, T. Takahashi, M. Hashimoto, *et al.*, IEEE Trans. Magn. **27**, 4736 (1991).

[22] W. F. Brown, Jr., Phys. Rev. **130**, 1677 (1963).

[23] S. H. Charap, P. L. Lu and Y. He, IEEE Trans. Magn. **33**, 978 (1997).

[24] D. Weller and A. Moser, IEEE Trans. Magn. 35, 4423 (1999).

[25] Y. Nakamura, J. Magn. Magn. Mat. **200**, 634 (1999).

[26] D. Weller, R. F. C. Farrow, R. F. Marks and G. R. Harp, Mat. Res. Soc. Symp. Proc. **313**, 791 (1993).

[27] K. Ouchi and N. Honda, IEEE Trans. Magn. **36**, 16 (2000).

[28] H. Katayama, S. Sawamura, Y. Ogimoto, *et al.*, J. Magn. Soc. Jpn. **23**, 233 (1999).

[29] J. J. M. Ruigrok, R. Coehoorn, S. R. Cumpson and H. W. Kesteren, J. Appl. Phys. **87**, 5398 (2000).

[30] R. M. H. New, R. F. W. Pease and R. L. White, Journal of Vacuum Science and Technology B **12**, 3196 (1994).

[31] S. Y. Chou, Proc. IEEE **85**, 652 (1997).

[32] J. Lohau, A. Moser, C. T. Rettner, M. E. Best and B. D. Terris, IEEE Trans. Magn. (preprint) (2000).

[33] S. Sun, C. B. Murray, D. Weller, L. Folks and A. Moser, it Science **287**, 1989 (2000).

[34] T. R. McGuire and R. I. Potter, IEEE Trans. Magn. MAG-**11**, 1018 (1975).

[35] P. L. Rossiter, *The electrical resistivity of metals and alloys* (Cambridge University Press, Cambridge, 1987).

[36] L. Neel, Ann. Phys. **2**, 61 (1967).

[37] S. S. P. Parkin, in *Ultrathin Magnetic Structures*, Vol. II, edited by B. Heinrich and J. A. C. Bland (Spinger-Verlag, Berlin, 1994), p. 148.

[38] A. Fert and P. Bruno, in *Ultrathin Magnetic Structures*, Vol. II, edited by B.Heinrich and J. A. C. Bland (Springer-Verlag, Berlin, 1994), p. 82.

[39] S. S. P. Parkin, in *Annual Review of Materials Science*, Vol. 25, edited by B. W. Wessels (Annual Reviews Inc., Palo Alto, 1995), p. 357.

[40] P. M. Levy, in *Solid State Physics*, Vol. 47, edited by H. Ehrenreich and D. Turnbull (Academic Press, New York, 1994), p. 367.

[41] S. S. P. Parkin, Z. G. Li and D. J. Smith, Appl. Phys. Lett. **58**, 2710 (1991a).

[42] C. L. Chien, in *Annu. Rev. Mater. Sci.*, Vol. 25, edited by B. W. Wessels (Annual Reviews Inc., Palo Alto, 1995), p. 129.

[43] S. S. P. Parkin, A. Modak and D. J. Smith, Phys. Rev. B **47**, 9136 (1993).

[44] S. K. J. Lenczowski, M. A. M. Gijs, J. B. Giesbers, R. J. M. van de Veerdonk and W. J. M. de Jonge, Phys. Rev. B **50**, 9982 (1994).

[45] S. S. P. Parkin, N. More and K. P. Roche, Phys. Rev. Lett. **64**, 2304 (1990).

[46] S. S. P. Parkin, R. Bhadra and K. P. Roche, Phys. Rev. Lett. **66**, 2152 (1991b).

[47] S. S. P. Parkin, Mat. Fys. Medd. Dan. Vid. Selsk. **45**, 113 (1997).

[48] D. T. Pierce, J. Unguris and R. J. Celotta, in *Ultrathin Magnetic Structures*, Vol. II, edited by B. Heinrich and J. A. C. Bland (Spinger-Verlag, Berlin, 1994).

[49] S. S. P. Parkin, Phys. Rev. Lett. **67**, 3598 (1991).

[50] J. Unguris, R. J. Celotta and D. T. Pierce, Phys. Rev. Lett. **67**, 140 (1991).

[51] W. Weber, Europhys. Lett. **31**, 491 (1995).

[52] P. Bruno and C. Chappert, Phys. Rev. Lett. **67**, 1602 (1991).

[53] J. Mathon, M. Villeret, D. M. Edwards and R. B Muniz, J. Magn. Magn. Mat. **121**, 242 (1993).

[54] B. Dieny, V. S. Speriosu, S. S. P. Parkin, B. A. Gurney, D. R. Wilhoit and D. Mauri, Phys. Rev. B **43**, 1297 (1991).

[55] J. Nogues and I. K. Schuller, J. Magn. Magn. Mat. **192**, 203 (1999).

[56] A. E. Berkowitz and K. Takano, J. Magn. Magn. Mat. **200**, 552 (1999).

[57] A. Yelon, in *Physics of Thin Films: Advances in Research and Development*, Vol. 6, edited by M. Francombe and R. Hoffman (Academic Press, New York, 1971), p. 205.

[58] A. P. Malozemoff, J. Appl. Phys. **63**, 3874 (1988).

[59] N. C. Koon, Phys. Rev. Lett. **78**, 4865 (1997).

[60] W. F. Egelhoff, T. Ha, R. D. K. Misra, Y. Kadmon, J. Nir, C. J. Powell, M. D. Stiles, R. D. McMichael, C.-L. Lin, J. M. Sivertsen, J. H. Judy, K. Takano, A. E. Berkowitz, T. C. Anthony and J. A. Brug, J. Appl. Phys. **78**, 273 (1995).

[61] W. F. Egelhoff, P. J. Chen, C. J. Powell, M. D. Stiles, R. D. McMichael, J. H. Judy, K. Takano and A. E. Berkowitz, J. Appl. Phys. **82**, 6142 (1997).

[62] R. E. Camley and J. Barnas, Phys. Rev. Lett. **63**, 664 (1989).

[63] S. S. P. Parkin, Appl. Phys. Lett. **61**, 1358 (1992).

[64] S. S. P. Parkin, Phys. Rev. Lett. **71**, 1641 (1993).

[65] C. Hwang and F. J. Himpsel, Phys. Rev. B **52**, 15368 (1995).

[66] S. S. P. Parkin and D. Mauri, Phys. Rev. B **44**, 7131 (1991).

[67] S. S. P. Parkin and D. E. Heim, in *US Patent and Trademark Office patent #5,465,185* (IBM, USA, 1995).

[68] A. H. Eschenfelder, *Magnetic Bubble Technology* (Springer-Verlag, Berlin, 1980).

[69] W. Renwick and A. J. Cole, *Digital Storage Systems* (Chapman and Hall Ltd., London, 1971).

[70] B. D. Cullity, *Introduction to Magnetic Materials* (Addison-Wesley Publishing Company, Reading, 1972).

[71] J. M. Daughton, Thin Solid Films **216**, 162 (1992).

[72] D. D. Tang, P. K. Wang, V. S. Speriosu and S. Le, IEEE Trans. Magn. **31**, 3206 (1995).

[73] R. E. Scheuerlein, in *NCE Seventh Biennial IEEE International Nonvolatile Memory Technology Conference Proc. IEEE Inter. Nonvolatile Memory Technology Conf., June 22-24, 1998* (IEEE, Albuquerque, NM, USA, 1998), p. 47.

[74] W. J. Gallagher, S. S. P. Parkin, R. E. Scheuerlein, *et al.*, in *United States Patent and Trademark Office patent #5,640,343* (IBM, USA, 1997).

[75] J. M. Daughton, J. Appl. Phys. **81**, 3758 (1997).

[76] S. Gider, B.-U. Runge, A. C. Marley and S. S. P. Parkin, Science **281**, 797 (1998).

# Index